State Estimation in Chemometrics

State Estimation in Chemometrics

The Kalman Filter and Beyond

Second Edition

Pierre Thijssen

Sillas Hadjiloucas

WP

WOODHEAD
PUBLISHING

ELSEVIER   An imprint of Elsevier

Woodhead Publishing is an imprint of Elsevier
The Officers' Mess Business Centre, Royston Road, Duxford, CB22 4QH, United Kingdom
50 Hampshire Street, 5th Floor, Cambridge, MA 02139, United States
The Boulevard, Langford Lane, Kidlington, OX5 1GB, United Kingdom

Notices
Knowledge and best practice in this field are constantly changing. As new research and experience
broaden our understanding, changes in research methods, professional practices, or medical
treatment may become necessary.

Practitioners and researchers must always rely on their own experience and knowledge in evaluating
and using any information, methods, compounds, or experiments described herein. In using such
information or methods they should be mindful of their own safety and the safety of others,
including parties for whom they have a professional responsibility.

To the fullest extent of the law, neither the Publisher nor the authors, contributors, or editors,
assume any liability for any injury and/or damage to persons or property as a matter of products
liability, negligence or otherwise, or from any use or operation of any methods, products,
instructions, or ideas contained in the material herein.

Library of Congress Cataloging-in-Publication Data
A catalog record for this book is available from the Library of Congress

British Library Cataloguing-in-Publication Data
A catalogue record for this book is available from the British Library

ISBN: 978-0-08-102603-8 (print)
ISBN: 978-0-08-102622-9 (online)

For information on all Woodhead publications
visit our website at https://www.elsevier.com/books-and-journals

Publisher: Susan Dennis
Acquisitions Editor: Kathryn Eryilmaz
Editorial Project Manager: Mona Zahir
Production Project Manager: Sojan P. Pazhayattil
Cover Designer: Alan Studholme & Jim Wilkie

Typeset by SPi Global, India

Working together
to grow libraries in
developing countries

www.elsevier.com • www.bookaid.org

Inscription

To obtain wisdom,
remove things every day.
To attain knowledge,
add things every day;

Lao Tzu

Contents

About the authors

Dr. Pierre Cornelis Thijssen studied chemistry at the Radboud University Nijmegen in the Netherlands, obtaining his Master's Degree in 1978. He then moved to the University of Amsterdam in the Netherlands, where he graduated in 1986 with a PhD based on his thesis entitled "State Estimation in Chemometrics," which is the basis of this book. Since then, Dr. Thijssen has worked for various companies as a laboratory manager and chemometrician.

Dr. Sillas Hadjiloucas received his Master's Degree in Pure & Applied Biology in 1992 from the University of Leeds, United Kingdom, and in 1996, a PhD in Cybernetics from the University of Reading, United Kingdom. He worked on his Postdoc in Germany until the end of 1999. Dr. Hadjiloucas had worked as permanent member of Staff at Reading since January 2000, originally at the School of Systems Engineering (Cybernetics) and later transferred to the newly formed Bio-Engineering Department at Reading in Autumn 2016. His current appointment as Associate Professor is a joint one combining teaching and research.

Chapter 1

Introduction

Chapter outline

1.1 History

The development of data processing techniques can be reviewed briefly in a historical perspective. At the beginning of the 19th century, Gauss (1809) developed the method of least squares and employed it in a simple orbit measurement system. During the next hundred years, several others made contributions to the field of estimation. A breakthrough came when Fisher (1910), working with probability density functions, reinvented the approach of the maximum likelihood. Much of this work has been employed thereafter in the broad area of statistics. A major change of viewpoint occurred when Kolmogorov (1941) and Wiener (1942) operated on random processes in the frequency domain. This approach describes the estimation problem in terms of correlation functions and the filter impulse response. It was limited to stationary processes and ensures only optimal estimates in the steady-state regime. Over the next 20 years, this work was extended in an often-cumbersome way to include non-stationary and multiple sensor systems. In the early 1960s, Kalman et al. (1960) advanced estimation with the concept of the state space model in the time domain and set the foundation of modern state estimation.

State estimation is concerned with the extraction of noise from measurements about some quantities that are essential to a system. A state is a minimal set of values sufficient to describe the behavior of a system. Three types of estimation are of interest: prediction is concerned with extrapolation of the state into the future, filtering recovers the state using measurements up to the current point, and smoothing involves interpolation of the state backward in the past. The Kalman filter is probably the most common estimation technique used in practice. Here, prediction and filtering are combined for an optimal performance of the estimation procedure. This approach is based on the online or recursive, rather than the batch processing of the measurements. It is ideally suited for computer implementation in automated systems and meets a broad application range: from ship navigation, image enhancement, process control, satellite orbit tracking, aircraft autopilot, earthquake forecasting to water resource planning.

State Estimation in Chemometrics. https://doi.org/10.1016/B978-0-08-102603-8.00001-8

The present book governs particularly applications of state estimation in the field of chemometrics. A lot of problems may arise when state estimation is applied in the practice of chemometrics. State estimation, for example, does not solve either the problem of modeling, how to acquire the noise statistics, how to select an optimal measurement schedule, or how to deal with computational errors and so on. Other design criteria, in addition to those used to derive the estimation algorithms, must be imposed to resolve such requirements.

Therefore, the blending together of state estimation and chemometrics is shown to be fruitful for both approaches.

At a first glance, the investigated systems in chemometrics often behave nonlinearly and/or nonstationary, and the modeling problem first has to be tackled.

In this book, mainly discrete linear state space models and preset noise statistics are involved for state estimation. In addition, some extensions are made toward the application of nonlinear models and the adaptation of the noise variances.

1.2 Chemometrics

Chemical analysis is referred to as the qualitative and/or quantitative determination of unknown constituents in samples. Here, analytical chemistry is devoted to the use and development of methods to enable chemical analysis. In less than a century, analytical chemistry has been developed from a mystic art to a reliable science. Nowadays, chemical analysis offers an important contribution to many organizations in society. Applications can be found for example in the petrochemical industry, clinical health survey, food quality assurance, and environmental pollution control.

With regard to history, a number of major stages can be distinguished:

In the manual stage, the analyst carries out chemical analysis with common laboratory glasswork and tools. The analyst possibly with help of a balance, polarimeter, or densitometer performs the measurement visually. The manual methods allow for easy operations and are often inexpensive. However, in practice, they may become tedious and manpower consuming. Examples of the former are gravimetric analysis, color-indicated titrations, and test tube procedures.

The instrumental stage introduces a great variety of novelties based on chemical or physical effects, which are transformed into an electrical signal. The measurement is performed by means of a recorder, voltmeter, or oscilloscope. The calculation of the required results follows after measurement by the analyst with simple arithmetics and graphics. The first sign of this development can be traced back to the early previous century. The design of instrumental methods grew simultaneously with the progress made in electronics. Contributions can be found in spectroscopy, chromatography, electrochemistry, flow injection analysis, etc.

Recently, the digital computer became available as a new achievement. Chemical analysis can now be exploited in a more efficient way by the capability to store and to process large amount of data. The automated stage introduces the new avenue of chemometrics to achieve, maintain, and improve the quality (or precision, accuracy, time, costs) of the analytical results. Chemometrics investigates strategies in chemical analysis to obtain a maximum of relevant information with minimal means and efforts. Mathematical and statistical methods are applied to design or to select optimal procedures and experiments. Various examples of progress can be given: the description, control, and surveillance of time series; experimental optimization by factorial designs or the simplex method; method selection by measurability and information theory; signal enhancement through estimation techniques; principal component analysis, partial least squares, and curve resolution; classification with pattern recognition; and finally digital simulation of laboratory organizations.

Nowadays, most of the instruments involved in chemical analysis are computer compatible and automated for control purposes, signal registration, data processing, and report generation. Automated instruments exploiting chemometrical techniques and innovations from artificial intelligence are the present state of the art. In the last category, the application of expert systems, neural networks, genetic algorithms, and support vector machines is worth mentioning.

What is the object of this book? Traditionally, the measurements are collected batchwise and computations follow afterward. The advent of today's computers offers the interactive coupling with an analytical instrument. Now, online data processing schemes such as the Kalman filter can be applied. As soon as a new measurement is available, calculations are updated and its results may be used more effectively. Relatively little attention has been paid to the linking of state estimation and chemometrics. Chemometrics should not be considered as just an outgrowth of but rather as a new dimension added to analytical chemistry. The first important step is to focus on the projection of the great variety of manual, instrumental, and automated methods for chemical analysis to the chemometrical axis. From this viewpoint, chemical analysis depends exclusively on the multicomponent, calibration, and titration methods, or combinations hereof. The application of modern state estimation in chemometrics is therefore demonstrated for these important elementary methods.

Firstly, some aspects of multicomponent analysis as applied in spectroscopy are investigated. Especially, the optimal design problem and the adaptation of the unknown measurement noise variance are of interest. In addition, the extension of the state space model with stochastic drift allows for the compensation of an unknown disturbance spectrum in the measured sample.

Secondly, state estimation is demonstrated in the field of the linear calibration method suffering from drifting parameters. Theoretical considerations particularly on state space modeling, evaluation of unknowns, quality control, optimal design, and variance reduction are investigated and applied in practice.

Furthermore, nonlinear estimation is applied when the calibration graph has an exponential shape.

Thirdly, state estimation is employed in the titration method for determining its curve and derivatives. From the estimated state, the setpoint(s) and inflection point(s) can be evaluated offline afterward. The online control to obtain equidistant measurements using variable volume additions in the discrete titration and the online endpoint control in case of continuous titration are of particular interest.

Finally, online processing the measurements for multiple modeling, principal component analysis and the generalized standard addition method is described. Also, iterative target transformation factor analysis evaluated offline is outlined. It further features chapters on subspace identification methods, new applications and recent advances in chemometrics.

In the next section, an attempt is made to develop a modular framework for state estimation applied to the most elementary methods of chemical analysis. This approach should be a first step in chemometrics toward the development of intelligent analyzers.

1.3   System view

According to the *Oxford Illustrated Dictionary* (1975), a system is defined as "the complex whole, set of connected things or parts, organized body of material or immaterial things." The entire system for chemical analysis that has to be investigated can be divided into the hierarchical order: society, laboratory, analyst, computer, instrument, chemometrical and analytical methods. The laboratory routes the sample streams obtained from various sources in society toward the locations where the chemical analysis is carried out. Here, the analyst handles the samples with his skills, techniques, and equipment either manually or supported by instruments and computers. Finally, the required results are reported to the customer in some form.

This book deals particularly with the last sections encountered in the whole system: the chemometrical and analytical methods implemented in a computer and coupled with a specific instrument. Each of these subsystems has its specific control, processing, and communication devices. The characteristics of the devices will influence the design of the entire system. For example, the usually setup for a manual titration operates on color changes, which are percepted and controlled by the analyst. If a computer is integrated, proper changes may occur in the instrument. An automatic titrator provides online control, but most computations follow after the entire titration is finished. Nowadays, almost all automated analyzers still act with the computer as an offline calculation machine. Integration provides more than the summed performance of the single parts in the chemical analysis system. Online data processing techniques can integrate the computer in an optimal way. An example of such a technique is provided by state estimation. A state is a minimal set of variables whose numerical values are sufficient to describe the behavior of a system. In chemometrics, a state may refer to concentrations, sensitivities, unknown parameters in a function or even to a curve and its derivatives. The state space model separates two sets of equations: a system equation that models the dynamics in time, wavelength, or volume and a measurement equation that relates the observed output to the current state. By using the state space model and available measurements, one may estimate the state by recursive least squares or the Kalman filter.

A modular framework (Table 1.1) evolves based on the application of state estimation in chemometrics for the most elementary methods of chemical analysis, i.e., multicomponent, calibration, and titration. This framework runs as a red thread through all the following text and chapters and enables the development and construction of intelligent analyzers. A set of algorithms may be

TABLE 1.1 Modular framework for state estimation

Goal	Objective of the system, i.e., results of interest[a]
Modeling	The mathematical structure to be used; here a state space model
Initialization	Acquisition of a priori information, e.g., starting values
Prediction	Forward extrapolation of the estimated state in the consecutive variable
Retrodiction	Backward extrapolation of the estimated state in the consecutive variable
Filtering	Online improvement of the estimated state by new measurements
Smoothing	Offline improvement of the filtered and predicted state
Control	Actions to be taken with respect to the goal of the system
Adaptation	Acquisition of the noise variances used for estimation
Restriction	Correction of the estimate given by physical and/or chemical constraints, equalities
Evaluation	Transformation of the measurements and/or the estimated state to the required result
Optimization	The best experimental design has to be chosen online or offline
Verification	Checking the measurement performance with respect to the model used
Selection	Determination of the model, estimation techniques, or data preprocessing procedures to be implemented
Simulation	Mimicking the system and required measurements
Normality	Checking the normality of the innovations or residuals

[a]Most generally: to obtain with minimal means and efforts a maximum of relevant information with respect to the qualitative and/or quantitative determination of chemical compounds.

selected and combined to act optimally within a given system for chemical analysis. The computer is incorporated in the instrument to perform the processing, control, and communication tasks. The degree of operation with minimal human intervention and maximal efficiency depends on the level of automation. A description of the desired goal and a detailed knowledge of the model to be used for chemical analysis should be investigated first. Then, the selected algorithms for state estimation are computer implemented and linked to the communication and instrumental devices. Processing measurements and controlling the behavior accordingly can achieve the goal of the system. It should be noted that the nomenclature is by no means complete and definitive. Analogies and extensions to other data processing schemes commonly employed in chemometrics have to be developed and are still under investigation.

Chapter 2

Classical estimation

Chapter outline

2.1 Modeling

Many analytical relations are adequately described by means of a mathematical model that is linear and additive:

$$z = h_1 x_1 + h_2 x_2 + \cdots + h_n x_n + v \tag{2.1}$$

The contribution of each of the parameters x_i, $i = 1,2,\ldots,n$ is weighted by the coefficients h_i with the noise term v added to result in the measured signal z.

It is obvious that for the determination of the unknown parameters several equations are required. Thus, Eq. (2.1) can be written more concisely as:

$$z(k) = \mathbf{h}^{\mathrm{T}}(k)\mathbf{x} + v(k) \quad k = 1,2,\ldots,m \tag{2.2}$$

where $z(k)$ is the scalar measurement, $\mathbf{x}$ the n-column vector with parameters x_i, $\mathbf{h}^{\mathrm{T}}(k)$ an n-row vector, the transpose of $\mathbf{h}(k)$ with coefficients $h_i(k)$ and $v(k)$ the scalar noise term, k is an index denoting the kth equation.

To determine the x_i values, the number of equations m must be at least as large as the number of unknown parameters n, i.e., $m \geq n$.

The linear model is designed as a block diagram shown in Fig. 2.1.

Note that the parameter vector $\mathbf{x}$ remains constant as a function of the index k.

The equations together can be summarized in a single matrix-vector notation:

$$Z(m) = H(m)\mathbf{x} + V(m) \tag{2.3}$$

where $Z(m)$ and $V(m)$ are m-column vectors containing the measurements $z(k)$ and noise terms $v(k)$, respectively, for $k = 1,2,\ldots,m$ and $H(m)$ is the $m*n$ design matrix with rows $\mathbf{h}^{\mathrm{T}}(k)$. The design matrix $H(m)$ or the set $\{\mathbf{h}(k)\}$, $k = 1,2,\ldots,m$

State Estimation in Chemometrics. https://doi.org/10.1016/B978-0-08-102603-8.00002-X

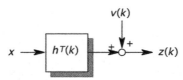

FIG. 2.1 Linear model for $k = 1,2,...,m$.

represents the combination of coefficients, factors, or instrumental variables. Each n-row design vector $\mathbf{h}^T(k)$ corresponds to a measurement, response, or experimental result $z(k)$ with the noise term $v(k)$.

2.2 Least squares

Given the measurements $Z(m)$ and design matrix $H(m)$, the basic problem is to find an estimate of the parameter vector $\mathbf{x}$. The solution to this problem depends on the assumptions made: the structure of the mathematical model, a priori statistical knowledge, and a suitable optimization criterion. One implicit assumption has already been made. The model structure is linear, which means that in Eq. (2.1) the measurement z, coefficients h_i, and the noise term v contain no parameters x_i.

In addition, the design matrix $H(m)$ is predetermined, known in advance or obtained by some standard procedure. $H(m)$ has maximal rank n, i.e., a proper solution from the given equations for the determination of the unknown parameters exists. It will further be assumed that the noise terms $v(k)$; $k = 1,2,...,m$ are not serially correlated, are independent of the other quantities involved, and are normally distributed with zero mean and given variance $R(k)$.

Because of the noise, it is not possible to determine the unknown parameters directly. However, as there are more measurements than unknowns, an estimator $\hat{\mathbf{x}}(m)$ of $\mathbf{x}$ that minimizes in some arbitrary sense the effect of the experimental errors $e(k)$ can be developed:

$$e(k) = z(k) - \mathbf{h}^T(k)\hat{\mathbf{x}}(m) \quad k = 1,2,...,m \tag{2.4}$$

For classical least squares, the estimator is chosen to minimize the weighted sum of squared errors as a cost criterion:

$$J(m) = \{Z(m) - H(m)\hat{\mathbf{x}}(m)\}^T W(m)\{Z(m) - H(m)\hat{\mathbf{x}}(m)\} \tag{2.5}$$

We now assume that the $m*m$ weighting matrix $W(m)$ is symmetric and positive definite. If we wish equal weighting of the residuals, we simply let $W(m)$ be the identity matrix. The estimator of the unknowns $\hat{\mathbf{x}}(m)$ is an n-column vector.

The least squares estimator $\hat{\mathbf{x}}(m)$ can be determined by forming the first partial derivative of the cost criterion $J(m)$ with respect to $\hat{\mathbf{x}}(m)$, equate it to zero, and solve it to get the required estimate.

The matrix-vector differentiation formulas to be used here are:

$$\frac{d(\mathbf{a}^T\mathbf{x})}{d\mathbf{x}} = \frac{d(\mathbf{x}^T\mathbf{a})}{d\mathbf{x}} = \mathbf{a}^T \quad \frac{d(\mathbf{x}^T A\mathbf{x})}{d\mathbf{x}} = \mathbf{x}^T\left(A+A^T\right) \quad \frac{d^2(\mathbf{x}^T A\mathbf{x})}{d\mathbf{x}^2} = A+A^T \quad (2.6)$$

where $\mathbf{x}$ and $\mathbf{a}$ are n-column vectors and A is a square $n*n$ matrix.

From Eqs. (2.5) and (2.6) and taking the transpose thereafter, it follows that:

$$\left(\frac{dJ(m)}{d\hat{\mathbf{x}}(m)}\right)^T = -2H^T(m)W(m)\{Z(m) - H(m)\hat{\mathbf{x}}(m)\} = 0 \qquad (2.7)$$

which gives directly the solution of the least squares problem to be used for filtering:

$$\hat{\mathbf{x}}(m) = \left\{H^T(m)W(m)H(m)\right\}^{-1}H^T(m)W(m)Z(m) \qquad (2.8)$$

Note that the second partial derivative is:

$$\frac{d^2J(m)}{d\hat{\mathbf{x}}(m)^2} = 2H^T(m)W(m)H(m) \qquad (2.9)$$

This matrix is positive definite as long as $H(m)$ has maximal rank n and $W(m)$ is symmetric and positive definite, so solution (2.8) is unique and minimizes $J(m)$.

When the weighting matrix $W(m)$ is chosen as the inverse of the noise covariance matrix $\mathbf{R}(m)$, the estimate is in statistical sense the best linear unbiased estimate with minimum error covariance:

$$\hat{\mathbf{x}}(m) = \left\{H^T(m)\mathbf{R}^{-1}(m)H(m)\right\}^{-1}H^T(m)\mathbf{R}^{-1}(m)Z(m) \qquad (2.10)$$

$$\text{or: } \hat{\mathbf{x}}(m) = \left\{\sum_{k=1}^{m}\mathbf{h}(k)R^{-1}(k)\mathbf{h}^T(k)\right\}^{-1}\sum_{k=1}^{m}\mathbf{h}(k)R^{-1}(k)z(k) \qquad (2.11)$$

where $\mathbf{R}(m)$ is an $m*m$ diagonal matrix with the noise variances $R(k)$; $k=1,2,\ldots,m$ on the diagonal. The $n*n$ covariance matrix $P(m)$ is defined by:

$$P(m) = \left\{H^T(m)\mathbf{R}^{-1}(m)H(m)\right\}^{-1} = \left\{\sum_{k=1}^{m}\mathbf{h}(k)R^{-1}(k)\mathbf{h}^T(k)\right\}^{-1} \qquad (2.12)$$

If the noise variances $R(k)$ are all equal and setting $W(m)$ in Eq. (2.8) to the identity matrix unbiased adaptation of the constant noise variance $\hat{R}(m)$ is given by:

$$\hat{R}(m) = \frac{1}{m-n}[Z(m) - H(m)\hat{\mathbf{x}}(m)]^T[Z(m) - H(m)\hat{\mathbf{x}}(m)] \qquad (2.13)$$

$$\text{or: } \hat{R}(m) = \frac{1}{m-n}\sum_{k=1}^{m}[z(k) - \mathbf{h}^T(k)\hat{\mathbf{x}}(m)]^2 \qquad (2.14)$$

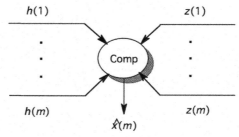

FIG. 2.2 Least squares estimation.

The covariance matrix $P(m)$ in case of adaptation is then given by:

$$P(m) = \hat{R}(m)\left[H^{\mathrm{T}}(m)H(m)\right]^{-1} \tag{2.15}$$

It should be noted that in classical least squares all the measurements have to be collected first to perform the estimation afterward and a growing memory requirement is inherent to this offline scheme.

The least squares estimation is depicted as a block diagram in Fig. 2.2.

For evaluation, the confidence interval for the individual x_i of $\mathbf{x}$ follows from the element $\hat{x}_i(m)$ of the estimate $\hat{\mathbf{x}}(m)$ and the diagonal elements $P_{ii}(m)$ of the covariance matrix $P(m)$:

$$\hat{x}_i(m) \pm t(1 - \alpha/2, m - n)\sqrt{P_{ii}(m)} \tag{2.16}$$

where $t(1 - \alpha/2, m - n)$ is the Student's t value for a given confidence level α and $m - n$ degrees of freedom.

A confidence interval for the noise variance R can also be derived:

$$\frac{m-n}{\chi^2(1 - \alpha/2, m - n)}\hat{R}(m) \leq R \leq \frac{m-n}{\chi^2(\alpha/2, m - n)}\hat{R}(m) \tag{2.17}$$

where $\chi^2(1 - \alpha/2, m - n)$ and $\chi^2(\alpha/2, m - n)$ are the chi-square values for a given confidence level α and $m - n$ degrees of freedom.

Critical table values of Student's t and chi-square can be found in statistical handbooks.

2.3 Curve fitting

We have examined the least squares problem for a linear model during the previous section. A linear model is often too restrictive and it is useful to consider the following, more general nonlinear model. Suppose the unknown $\mathbf{x}$ is related to the measurements $Z(m)$ according to:

$$Z(m) = \mathbf{g}(\mathbf{x}, m) + V(m) \tag{2.18}$$

where the m-column vector $\mathbf{g}(\mathbf{x}, m)$ represents some nonlinear function of $\mathbf{x}$.

The least squares estimator $\hat{\mathbf{x}}(m)$ of $\mathbf{x}$ given $Z(m)$ is defined as the one that minimizes the cost criterion:

$$J(m) = [Z(m) - \mathbf{g}(\hat{\mathbf{x}}(m), m)]^{\mathrm{T}} W(m)[Z(m) - \mathbf{g}(\hat{\mathbf{x}}(m), m)] \qquad (2.19)$$

where $W(m)$ is again an $m*m$ weighting matrix, which is assumed to be symmetric and positive definite.

To determine the least squares estimate $\hat{\mathbf{x}}(m)$, take the first partial derivative of the cost criterion $J(m)$ with respect to $\hat{\mathbf{x}}(m)$, equate it to zero, take the transpose, and solve it:

$$\left(\frac{dJ(m)}{d\hat{\mathbf{x}}(m)}\right)^{\mathrm{T}} = -2\left(\frac{d\mathbf{g}(\hat{\mathbf{x}}(m), m)}{d\hat{\mathbf{x}}(m)}\right)^{\mathrm{T}} W(m)\{Z(m) - \mathbf{g}(\hat{\mathbf{x}}(m), m)\} = 0 \qquad (2.20)$$

Note that when $\mathbf{g}(\hat{\mathbf{x}}(m), m) = H(m)\hat{\mathbf{x}}(m)$ and thus $\frac{d\mathbf{g}(\hat{\mathbf{x}}(m), m)}{d\hat{\mathbf{x}}(m)} = H(m)$ Eq. (2.20) reduces to Eq. (2.7) with its unique solution (2.8). However, when $\mathbf{g}(\hat{\mathbf{x}}(m), m)$ is nonlinear, (2.20) represents a set of nonlinear equations for which in general no closed form solution exists.

Now, suppose that an initial estimate $\hat{\mathbf{x}}^*(m)$ of $\mathbf{x}$ is available. Using $\hat{\mathbf{x}}^*(m)$ expand $\mathbf{g}(\mathbf{x}, m)$ in Eq. (2.18) in a Taylor series:

$$\mathbf{g}(\mathbf{x}, m) = \mathbf{g}(\hat{\mathbf{x}}^*(m), m) + \frac{d\mathbf{g}(\hat{\mathbf{x}}^*(m), m)}{d\hat{\mathbf{x}}^*(m)}\{\mathbf{x} - \hat{\mathbf{x}}^*(m)\} + h.o.t. \qquad (2.21)$$

The higher-order terms ($h.o.t.$) are neglected. Now define:

$$H(m) \text{ with rows } \mathbf{h}^{\mathrm{T}}(k) = \left.\frac{dg(\mathbf{x}, k)}{d\mathbf{x}}\right|_{\mathbf{x}=\hat{\mathbf{x}}^*(m)} \qquad k = 1, 2, \ldots, m \qquad (2.22)$$

Using (2.21) and (2.22) in (2.20) leads after some manipulations to a new approximate of the least squares estimate as the following equation for filtering:

$$\hat{\mathbf{x}}(m) = \hat{\mathbf{x}}^*(m) + \left\{H^{\mathrm{T}}(m)W(m)H(m)\right\}^{-1} H^{\mathrm{T}}(m)W(m)\{Z(m) - \mathbf{g}(\hat{\mathbf{x}}^*(m), m)\} \qquad (2.23)$$

Thus, this gives an iterative procedure for refining the estimate. After each computation, $\hat{\mathbf{x}}(m)$ can be replaced by $\hat{\mathbf{x}}^*(m)$ and the next cycle of linearization is repeated until this ends somewhere.

When the weighting matrix $W(m)$ is chosen as the inverse of the noise covariance matrix $\mathbf{R}(m)$, then iteration generates the least squares estimate:

$$\hat{\mathbf{x}}(m) = \hat{\mathbf{x}}^*(m) + \left\{ H^\mathrm{T}(m)\mathbf{R}^{-1}(m)H(m) \right\}^{-1} H^\mathrm{T}(m)\mathbf{R}^{-1}(m)\{Z(m) - \mathbf{g}(\hat{\mathbf{x}}^*(m), m)\} \tag{2.24}$$

If the weighting matrix $W(m)$ in (2.23) is assumed to be the identity matrix and $\mathbf{R}(m)$ is an $m*m$ diagonal matrix with the diagonal noise variances $R(k)$; $k = 1,2,...,m$ all equal adaptation of the constant noise variance $\hat{R}(m)$ is given by:

$$\hat{R}(m) = \frac{1}{m-n}[Z(m) - \mathbf{g}(\hat{\mathbf{x}}(m), m)]^\mathrm{T}[Z(m) - \mathbf{g}(\hat{\mathbf{x}}(m), m)] \tag{2.25}$$

$$\text{or}: \quad \hat{R}(m) = \frac{1}{m-n}\sum_{k=1}^{m}[z(k) - g(\hat{\mathbf{x}}(m), k)]^2$$

The covariance matrix $P(m)$ in Eqs. (2.12) and (2.15), the confidence interval for both the unknown x_i in Eq. (2.16), and the noise variance R in Eq. (2.17) are the same in the nonlinear case.

For termination of the iteration process, one may consider as a stopping criterion:

$$\left| \frac{\hat{R}^*(m) - \hat{R}(m)}{\hat{R}^*(m)} \right| < \delta \tag{2.26}$$

where "$|\ |$" denotes the absolute value and δ is a small value to be chosen.

As an alternative form for the analytical solution (2.22) of the derivatives of $\mathbf{g}(\hat{\mathbf{x}}^*(m), m)$, one can use as approximation for the coefficients $h_i(k)$ of $H(m)$:

$$h_i(k) = \frac{dg(\hat{\mathbf{x}}^*(m), k)}{d\hat{x}^*_i(m)} \approx \frac{g(\hat{\mathbf{x}}^*(m), k)_{\hat{x}^*_i(m)+\Delta} - g(\hat{\mathbf{x}}^*(m), k)_{\hat{x}^*_i(m)-\Delta}}{2\Delta} \tag{2.27}$$

for $k = 1,2,...,m$ and $i = 1,2,...,n$; where Δ is a small value to be chosen.

There are various other algorithms possible for curve fitting, which are not discussed here. The one presented here is justified because of its simplicity and link to nonlinear state estimation.

2.4 Recursive approach

In order to avoid a growing memory filter, as in classical least squares processing the measurements offline, an estimate is sought in a recursive form. We switch from final index m to k because we are interested in estimates at k between the measurements sequencing from 1 to m, which allows determining the results online.

What now follows is a bunch of matrix manipulations in order to derive the recursive least squares algorithm and simultaneously obtain some other valuable matrix equations that can be used later for state estimation.

The parameter vector $\mathbf{x}$ is estimated in such a way that the estimator $\hat{\mathbf{x}}(k)$ (2.10) minimizes the cost criterion $J(k)$ (2.5). Redefining (2.11) with use of (2.12) gives:

$$\hat{\mathbf{x}}(k) = P(k)\mathbf{b}(k) \tag{2.28}$$

where $P(k)^{-1} = \sum_{i=1}^{k}\mathbf{h}(i)R^{-1}(i)\mathbf{h}^{T}(i)$ and $\mathbf{b}(k) = \sum_{i=1}^{k}\mathbf{h}(i)R^{-1}(i)z(i)$.

This results in the analogous recursive relations:

$$P^{-1}(k) = P^{-1}(k-1) + \mathbf{h}(k)R^{-1}(k)\mathbf{h}^{T}(k) \tag{2.29}$$

$$\mathbf{b}(k) = \mathbf{b}(k-1) + \mathbf{h}(k)R^{-1}(k)z(k)$$

Some rewriting of (2.28) gives:

$$\hat{\mathbf{x}}(k) = P(k)\mathbf{b}(k) = P(k)\left[\mathbf{b}(k-1) + \mathbf{h}(k)R^{-1}(k)z(k)\right] \tag{2.30}$$

$$= P(k)\left[P^{-1}(k-1)\hat{\mathbf{x}}(k-1) + \mathbf{h}(k)R^{-1}(k)z(k)\right]$$

$$= P(k)\left[\{P^{-1}(k) - \mathbf{h}(k)R^{-1}(k)\mathbf{h}^{T}(k)\}\hat{\mathbf{x}}(k-1) + \mathbf{h}(k)R^{-1}(k)z(k)\right]$$

$$= \hat{\mathbf{x}}(k-1) + P(k)\mathbf{h}(k)R^{-1}(k)\{z(k) - \mathbf{h}^{T}(k)\hat{\mathbf{x}}(k-1)\}$$

$$\text{or}: \quad \hat{\mathbf{x}}(k) = \hat{\mathbf{x}}(k-1) + \mathbf{k}(k)\{z(k) - \mathbf{h}^{T}(k)\hat{\mathbf{x}}(k-1)\} \tag{2.31}$$

$$\mathbf{k}(k) = P(k)\mathbf{h}(k)R^{-1}(k) \tag{2.32}$$

where $\mathbf{k}(k)$ is the n-column gain vector. Rewriting:

$$P(k-1) = P(k-1)P^{-1}(k)P(k) \tag{2.33}$$

$$= P(k-1)\left[P^{-1}(k-1) + \mathbf{h}(k)R^{-1}(k)\mathbf{h}^{T}(k)\right]P(k)$$

$$= P(k) + P(k-1)\mathbf{h}(k)R^{-1}(k)\mathbf{h}^{T}(k)P(k)$$

Multiplying the left sides of the equation both with $\mathbf{h}^{T}(k)$ gives:

$$\mathbf{h}^{T}(k)P(k-1) = \left\{1 + \mathbf{h}^{T}(k)P(k-1)\mathbf{h}(k)R^{-1}(k)\right\}\mathbf{h}^{T}(k)P(k) \tag{2.34}$$

Without proof, the following formula for the determinant of the covariance matrix has shown to be valid and known as the matrix determinant lemma:

$$|P(k-1)| = \left\{\mathbf{h}^{T}(k)P(k-1)\mathbf{h}(k)/R(k) + 1\right\}|P(k)| \tag{2.35}$$

The matrix determinant lemma is useful later for optimization of the experimental design or as a divergence test.

Going further, substituting (2.34) into (2.33), we find:

$$P(k-1) = P(k) + P(k-1)\mathbf{h}(k)R^{-1}(k)\left\{1 + \mathbf{h}^{T}(k)P(k-1)\mathbf{h}(k)R^{-1}(k)\right\}^{-1}\mathbf{h}^{T}(k)P(k-1)$$

$$\text{or}: \quad P(k) = P(k-1) - P(k-1)\mathbf{h}(k)\left\{\mathbf{h}^{T}(k)P(k-1)\mathbf{h}(k) + R(k)\right\}^{-1}\mathbf{h}^{T}(k)P(k-1) \tag{2.36}$$

Reformulating gives the matrix inversion lemma relating Eq. (2.29) with (2.36):

$$P(k) = P(k-1) - \mathbf{k}(k)\mathbf{h}^{\mathrm{T}}(k)P(k-1) \tag{2.37}$$

$$\mathbf{k}(k) = P(k-1)\mathbf{h}(k)\big\{\mathbf{h}^{\mathrm{T}}(k)P(k-1)\mathbf{h}(k) + R(k)\big\}^{-1} \tag{2.38}$$

The term between curly brackets in Eq. (2.38) is a scalar; hence, no matrix inversion is needed anymore in the computations.

Both expressions for the gain vector $\mathbf{k}(k)$ (2.32) and (2.38) can be shown to be equivalent. Filling in Eq. (2.32), the result (2.36) for the covariance matrix $P(k)$:

$$\mathbf{k}(k) = P(k)\mathbf{h}(k)R^{-1}(k) \tag{2.32}$$

$$= P(k-1)\mathbf{h}(k)\Big[1 - \big\{\mathbf{h}^{\mathrm{T}}(k)P(k-1)\mathbf{h}(k) + R(k)\big\}^{-1}\mathbf{h}^{\mathrm{T}}(k)P(k-1)\mathbf{h}(k)\Big]R^{-1}(k)$$

$$= P(k-1)\mathbf{h}(k)\big\{\mathbf{h}^{\mathrm{T}}(k)P(k-1)\mathbf{h}(k) + R(k)\big\}^{-1}$$
$$\cdot \big[\mathbf{h}^{\mathrm{T}}(k)P(k-1)\mathbf{h}(k) + R(k) - \mathbf{h}^{\mathrm{T}}(k)P(k-1)\mathbf{h}(k)\big]R^{-1}(k)$$

$$= P(k-1)\mathbf{h}(k)\big\{\mathbf{h}^{\mathrm{T}}(k)P(k-1)\mathbf{h}(k) + R(k)\big\}^{-1} = \mathbf{k}(k) \tag{2.38}$$

To summarize the recursive least squares algorithm for filtering:

$$\hat{\mathbf{x}}(k) = \hat{\mathbf{x}}(k-1) + \mathbf{k}(k)\big\{z(k) - \mathbf{h}^{\mathrm{T}}(k)\hat{\mathbf{x}}(k-1)\big\} \tag{2.39}$$

$$P(k) = P(k-1) - \mathbf{k}(k)\mathbf{h}^{\mathrm{T}}(k)P(k-1) \tag{2.40}$$

$$\mathbf{k}(k) = P(k-1)\mathbf{h}(k)\big\{\mathbf{h}^{\mathrm{T}}(k)P(k-1)\mathbf{h}(k) + R(k)\big\}^{-1} \tag{2.41}$$

An alternative expression for the covariance matrix $P(k)$ considers:

$$P(k) = \big[I - \mathbf{k}(k)\mathbf{h}^{\mathrm{T}}(k)\big]P(k-1)\big[I - \mathbf{k}(k)\mathbf{h}^{\mathrm{T}}(k)\big]^{\mathrm{T}} + \mathbf{k}(k)R(k)\mathbf{k}^{\mathrm{T}}(k) \tag{2.42}$$

$$= P(k-1) - \mathbf{k}(k)\mathbf{h}^{\mathrm{T}}(k)P(k-1) - P(k-1)\mathbf{h}(k)\mathbf{k}^{\mathrm{T}}(k)$$
$$+ \mathbf{k}(k)\big\{\mathbf{h}^{\mathrm{T}}(k)P(k-1)\mathbf{h}(k) + R(k)\big\}\mathbf{k}^{\mathrm{T}}(k) \tag{2.43}$$

Filling Eq. (2.41) once for the gain vector $\mathbf{k}(k)$ in the last term on the left side of (2.43) gives immediately Eq. (2.40).

Numerical computations by (2.42) will be better conditioned, keeping both the symmetry and positive definiteness of the covariance matrix $P(k)$. If the covariance matrix becomes ill-conditioned, one may consider a similarity transformation and/or square root filtering in order to improve the computations numerically.

2.5 Examples

Example 2.1: Multicomponent analysis of overlapped spectra is directly based on the described linear model (2.3). In spectroscopy, the law of Lambert-Beer holds, providing a linear and additive measurement model. In this case, the unknown vector $\mathbf{x}$ contains the concentrations in the sample to be estimated. The design matrix $H(m)$ is composed of the molar absorptivities, which are wavelength dependent and different for each individual component. These values are usually obtained by measuring the spectra of the pure components of interest. Each measurement $z(k)$ corresponds to the absorbance of the sample at a given wavelength k.

As example, consider the individual molar spectra for four components plus baseline depicted in Fig. 2.3. For simulation of the system, the concentrations involved are combined with the design matrix, whereafter Gaussian white noise with zero mean and variance $R(k) = 10^{-5}$ is added to obtain the measurements.

The measurements and their least squares fit are plotted in Fig. 2.4.

A similar approach can be applied to overlapped responses from other instrumental methods, when a linear and additive measurement model is adequate.

Example 2.2: One of the most frequent applications of classical least squares is probably the regression method. Here, the unknown parameters are determined from data that are assumed to follow a linear relationship. After calibration of the system by least squares regression, evaluation of unknown samples is performed.

As example, consider the quadratic regression model:

$$z(k) = a_2 c^2(k) + a_1 c(k) + a_0 \tag{2.44}$$

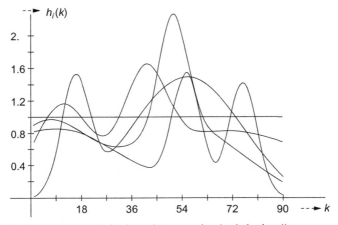

FIG. 2.3 Multicomponent analysis: the molar spectra involved plus baseline.

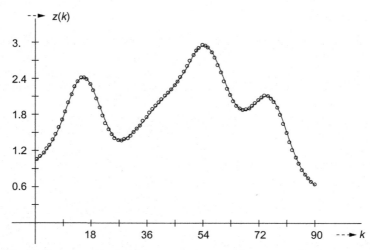

FIG. 2.4 Multicomponent analysis: the measurements and least squares fit.

In this case, the unknown sensitivities $\mathbf{x}^T = (a_2 \ a_1 \ a_0)$ have to be estimated.

The design matrix $H(m)$ contains the concentrations $c(k)$ at index $k = 1,2,\ldots,$ m in its rows $\mathbf{h}^T(k) = \left(c^2(k) \ c(k) \ 1 \right)$. A noise variance $R(k) = 10^{-4}$ is simulated.

In Fig. 2.5, the measurements $z(k)$ corrupted with noise, the least squares fit $\hat{z}(k)$, and its lower and upper 95% confidence limits $\underline{z}(k)$ and $\bar{z}(k)$ are shown:

$$\hat{z}(k) = \mathbf{h}^T(k)\hat{\mathbf{x}}(m) \tag{2.45}$$

$$\underline{z}(k), \bar{z}(k) = \hat{z}(k) \pm t(1 - \alpha/2, m - n)\left\{ \mathbf{h}^T(k)P(m)\mathbf{h}(k) + \hat{R}(m) \right\}^{1/2}$$

Example 2.3: Sometimes the regression follows an exponential shape by the nonlinear function:

$$g(\mathbf{x}, k) = x_\infty \left(1 - e^{-k_i c(k)} \right) + x_0 \tag{2.46}$$

where $c(k)$ is the concentration at index k, x_∞ the limiting value to infinity, k_i is the exponential constant, and x_0 is the baseline contribution. Here, the unknown parameters $\mathbf{x}^T = (x_\infty \ k_i \ x_0)$ have to be estimated in an iterative fashion by curve fitting from the measurements.

The design matrix $H(m)$ contains the derivatives in the rows for $k = 1,2,\ldots,m$:

$$\mathbf{h}^T(k) = \left(1 - e^{-k_i c(k)} \quad c(k)x_\infty e^{-k_i c(k)} \quad 1 \right) \tag{2.47}$$

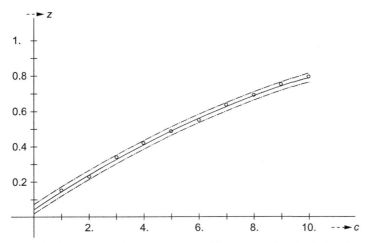

FIG. 2.5 Quadratic regression: the measurements and least squares fit. The *dashed lines* around represent the 95% confidence limits.

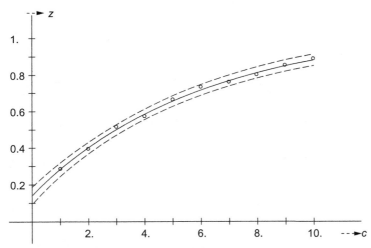

FIG. 2.6 Exponential regression: the measurements and final least squares fit after curve fitting. The *dashed lines* around represent the 95% confidence limits.

A noise variance $R(k) = 10^{-4}$ is used. In Fig. 2.6, the measurements $z(k)$ corrupted with noise, the least squares fit $\hat{z}(k)$, and its lower and upper 95% confidence limits $\underline{z}(k)$ and $\bar{z}(k)$ are shown:

$$\hat{z}(k) = g(\hat{\mathbf{x}}(m), k) \qquad (2.48)$$

$$\underline{z}(k), \overline{z}(k) = \hat{z}(k) \pm t(1 - \alpha/2, m - n)\{\mathbf{h}^T(k)P(m)\mathbf{h}(k) + \hat{R}(m)\}^{1/2}$$

Example 2.4: Within kinetics, the nonlinear model for a first-order consecutive reaction $A \rightarrow B \rightarrow C$ is described here. Analysis of chemical reaction profiles by monitoring the concentrations of one or more products and/or reactants can be accomplished by curve fitting. The measurement function that describes in time the dependence of the signal on the kinetic parameters, where only the final product is monitored, is given by:

$$g(\mathbf{x}, k) = x_\infty \left\{ 1 + \frac{k_1}{k_2 - k_1} e^{-k_2 t(k)} - \frac{k_2}{k_2 - k_1} e^{-k_1 t(k)} \right\} + x_0 \qquad (2.49)$$

where $t(k)$ is the time at index k, x_∞ the final value at infinity, k_i are reaction constants, and x_0 is the background contribution. The unknown parameter vector comprises $\mathbf{x}^T = (x_\infty \ k_1 \ k_2 \ x_0)$ and the design matrix $H(m)$ contains the derivatives in the rows $\mathbf{h}^T(k)$ for $k = 1, 2, \dots, m$:

$$\frac{dg(\mathbf{x}, k)}{dx_0} = 1 \quad \frac{dg(\mathbf{x}, k)}{dx_\infty} = 1 + \frac{k_1}{k_2 - k_1} e^{-k_2 t(k)} - \frac{k_2}{k_2 - k_1} e^{-k_1 t(k)} \qquad (2.50)$$

$$\frac{dg(\mathbf{x}, k)}{dk_1} = \frac{k_2 x_\infty}{k_2 - k_1} \left\{ t(k) e^{-k_1 t(k)} + \frac{1}{k_2 - k_1} \left(e^{-k_2 t(k)} - e^{-k_1 t(k)} \right) \right\}$$

$$\frac{dg(\mathbf{x}, k)}{dk_2} = \frac{k_1 x_\infty}{k_2 - k_1} \left\{ -t(k) e^{-k_2 t(k)} + \frac{1}{k_2 - k_1} \left(e^{-k_1 t(k)} - e^{-k_2 t(k)} \right) \right\}$$

In Fig. 2.7, the measurements with noise and initial fit before curve fitting are given. The measurements and final least squares fit after curve fitting are shown in Fig. 2.8. Here, a noise variance $R(k) = 10^{-5}$ is employed.

Example 2.5: In chromatography, the mathematical model to describe in time an elution peak is given by the Fraser-Suzuki equation:

$$g(\mathbf{x}, k) = H e^{-\left[(\ln 2)/A^2 \right] \left[\ln \left(1 + \frac{A(t(k) - t_r)}{\sigma \sqrt{2 \ln 2}} \right) \right]^2} \qquad (2.51)$$

where $t(k)$ is the time at index k, H the peak height, A the asymmetry factor, σ the standard deviation, and t_r the retention time of the peak. Furthermore, there are the conditions $t(k) > t_r - \sigma\sqrt{2 \ln 2}/A$ for $A > 0$ and $t(k) < t_r - \sigma\sqrt{2 \ln 2}/A$ for $A < 0$.

The baseline involved is a quadratic polynomial $b = a_0 + a_1 t(k) + a_2 t^2(k)$.

The unknown parameter vector comprises $\mathbf{x}^T = (H \ A \ t_r \ \sigma \ a_0 \ a_1 \ a_2)$.

Treatment of the derivatives can be simplified by redefining: $K = \sqrt{2 \ln 2}$, $S = 1 + A(t(k) - t_r)/(\sigma K)$ and $F = e^{-[(\ln 2)/A^2][\ln S]^2}$. Now, the design matrix $H(m)$ contains the derivatives in the rows $\mathbf{h}^T(k)$ for $k = 1, 2, \dots, m$:

$$\frac{dg(\mathbf{x}, k)}{dH} = F \quad \frac{dg(\mathbf{x}, k)}{dt_r} = HFK \ln S / (A\sigma S) \qquad (2.52)$$

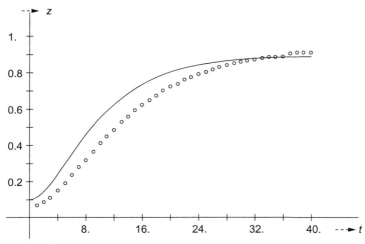

FIG. 2.7 Kinetics: the measurements and initial fit.

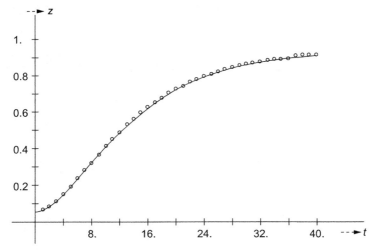

FIG. 2.8 Curve-fitting kinetics: the measurements and final least squares fit.

$$\frac{dg(\mathbf{x}, k)}{d\sigma} = \left[\frac{dg(\mathbf{x}, k)}{dt_r}\right](t(k) - t_r)/\sigma$$

$$\frac{dg(\mathbf{x}, k)}{dA} = \left[\frac{dg(\mathbf{x}, k)}{dt_r}\right][K\sigma S \ln S - A(t(k) - t_r)]/A^2$$

$$\frac{dg(\mathbf{x}, k)}{da_0} = 1 \quad \frac{dg(\mathbf{x}, k)}{da_1} = t(k) \quad \frac{dg(\mathbf{x}, k)}{da_2} = t^2(k)$$

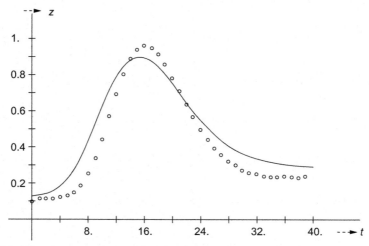

FIG. 2.9 Chromatography (one peak): the measurements and initial fit.

In Figs. 2.9 and 2.10, the measurements, initial fit, and the final least squares fit after curve fitting are depicted. The noise variance used for simulation is $R(k) = 10^{-5}$.

Example 2.6: A three-peak system with quadratic baseline in chromatography is considered. The vector of unknowns $\mathbf{x}$ is extended to cover the parameters of each individual peak and thereafter a polynomial baseline is added. If there are p peaks and the baseline is of degree q, the extended model requires $4p + q + 1$ parameters. For each individual peak, the derivatives to be used for

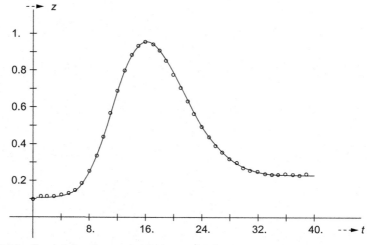

FIG. 2.10 Curve-fitting chromatography (one peak): the measurements and final least squares fit.

curve fitting are given in Example 2.5 by Eq. (2.52). For the three-peak system with quadratic baseline the measurements, initial fit and final least squares fit after curve fitting are plotted in Figs. 2.11 and 2.12. Here, 15 parameters are involved. The noise variance applied is $R(k) = 10^{-5}$.

Example 2.7: One classic example to be mentioned is found in spectroscopy where the measurements are described by a number of combined Gaussian and Lorentzian curves. Each individual peak is described by:

$$f = He^{-\frac{(x(k)-\bar{x})^2}{2s^2}} \quad \ell = \frac{A}{1 + \frac{(x(k)-\bar{x})^2}{2\ln 2s^2}} \tag{2.53}$$

where f is a Gaussian curve, ℓ is a Lorentzian curve, $x(k)$ is the wavelength at index k, $\bar{x}$ the mean value, s the standard deviation, H and A are peak heights.

The term ln2 in front of the squared standard deviation s^2 in the Lorentzian curve is found after making the peaks similar at heights $H/2$ and $A/2$.

Introducing the term $c = (x(k) - \bar{x})/s$, the derivatives for a Gaussian and Lorentzian curve are defined by:

$$\frac{df}{dH} = e^{-\frac{c^2}{2}} \qquad \frac{d\ell}{dA} = \frac{1}{1 + \frac{c^2}{2\ln 2}} \tag{2.54}$$

$$\frac{df}{d\bar{x}} = \left[\frac{df}{dH}\right] H \frac{c}{s} \qquad \frac{d\ell}{d\bar{x}} = \left[\frac{d\ell}{dA}\right]^2 A \frac{c}{\ln 2s}$$

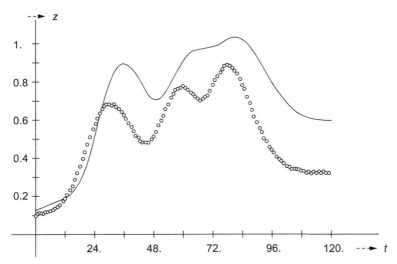

FIG. 2.11 Chromatography (three peaks): the measurements and initial fit.

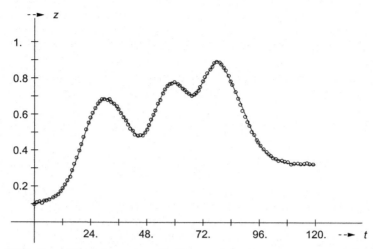

FIG. 2.12 Curve-fitting chromatography (three peaks): the measurements and final least squares fit.

$$\frac{df}{ds}=\left[\frac{df}{d\bar{x}}\right]c \qquad \frac{d\ell}{ds}=\left[\frac{d\ell}{d\bar{x}}\right]c$$

The baseline involved is a quadratic polynomial $b=a_0+a_1x(k)+a_2x^2(k)$.

A one-peak system can be described as a combination of a Gaussian and Lorentzian curve with the baseline added:

$$g(\mathbf{x},k)=f+\ell+b \qquad\qquad (2.55)$$

The unknown parameter vector comprises $\mathbf{x}^T=(H\ \bar{x}\ s\ A\ a_0\ a_1\ a_2)$.

The design matrix $H(m)$ contains the derivatives in the rows $\mathbf{h}^T(k)$ for $k=1,2,\ldots,m$:

$$\frac{dg(\mathbf{x},k)}{dH}=\frac{df}{dH} \qquad\qquad \frac{dg(\mathbf{x},k)}{dA}=\frac{d\ell}{dA} \qquad\qquad (2.56)$$

$$\frac{dg(\mathbf{x},k)}{d\bar{x}}=\frac{df}{d\bar{x}}+\frac{d\ell}{d\bar{x}} \qquad\qquad \frac{dg(\mathbf{x},k)}{ds}=\frac{df}{ds}+\frac{d\ell}{ds}$$

$$\frac{dg(\mathbf{x},k)}{da_0}=1 \qquad\qquad \frac{dg(\mathbf{x},k)}{da_1}=x(k) \qquad\qquad \frac{dg(\mathbf{x},k)}{da_2}=x^2(k)$$

For more peaks, the vector of unknowns $\mathbf{x}$ is extended to cover the parameters of each individual peak and thereafter a polynomial baseline is added. If there are p peaks and the baseline is of degree q, the extended model requires $4p+q+1$ parameters. For each individual peak, the derivatives to be used are given by Eqs. (2.54) and (2.56). As an example, a three-peak system with quadratic baseline in spectroscopy is considered, i.e., 15 parameters are involved. For the three-peak system the measurements, initial fit, and final least squares fit after curve fitting are plotted in Figs. 2.13 and 2.14. The noise variance applied is $R(k)=10^{-5}$.

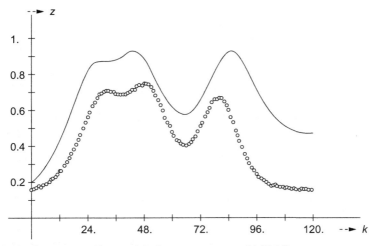

FIG. 2.13   Spectroscopy (three peaks): the measurements and initial fit.

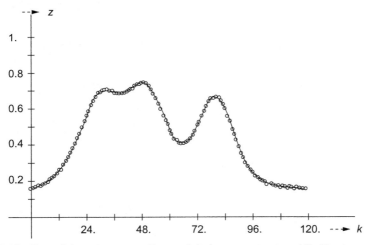

FIG. 2.14   Curve-fitting spectroscopy (three peaks): the measurements and final least squares fit.

Chapter 3

State estimation

Chapter outline

3.1 Modeling

The internal structure of a system can be described by the concept of the state, which is defined as a minimal set or vector of variables whose numerical values are sufficient to describe the behavior of a system. The physical or chemical variables are the inputs and outputs of the system that can be measured. A state may refer to variables such as concentrations, sensitivities, unknown parameters in a function or even to a curve and its derivatives. The state space approach is important, because it separates two sets of equations. The system equation (3.1) is a dynamic model that describes the behavior of the state in time, wavelength, or volume and can be a set of difference equations. The measurement equation (3.2) relates the observed output to the state of the system. Both equations may be subject to random disturbances.

A discrete linear state space model is described by:

$$\mathbf{x}(k) = F(k, k-1)\mathbf{x}(k-1) + \mathbf{w}(k-1) \tag{3.1}$$

$$z(k) = \mathbf{h}^{\mathrm{T}}(k)\mathbf{x}(k) + v(k) \tag{3.2}$$

The state space model is designed as a block diagram in Fig. 3.1. Here, k is an index denoting a sequence number and $\mathbf{x}(k-1)$ is the n-column state vector. The new state $\mathbf{x}(k)$ in the sequence is obtained by the product of the old state $\mathbf{x}(k-1)$ and the $n*n$ transition matrix $F(k,k-1)$ with addition of the n-column system noise vector $\mathbf{w}(k-1)$. The state $\mathbf{x}(k)$ is weighted by the transposed n-row measurement vector $\mathbf{h}^{\mathrm{T}}(k)$ and the scalar measurement noise $v(k)$ added to yield the scalar signal $z(k)$. The assumption of linearity of the state space

State Estimation in Chemometrics. https://doi.org/10.1016/B978-0-08-102603-8.00003-1

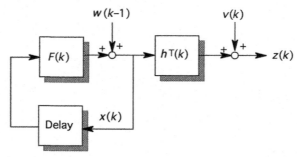

FIG. 3.1 State space model.

model implies that the transition matrix $F(k,k-1)$ and the measurement vector $\mathbf{h}^T(k)$ have to be known a priori.

The given state space model is not unique, because there are infinite possible representations. The various equivalent state descriptions can be interrelated by similarity transformations. Defining a new state vector $\mathbf{x}^*(k)$ by the relation $\mathbf{x}(k) = T(k)\mathbf{x}^*(k)$, with the similarity matrix $T(k)$ as a nonsingular transformation matrix, generates the alternative state space model:

$$\mathbf{x}^*(k) = T^{-1}(k)\mathbf{x}(k)$$
$$F^*(k,k-1) = T^{-1}(k)F(k,k-1)T(k-1) \tag{3.3}$$

$$\mathbf{w}^*(k-1) = T^{-1}(k)\mathbf{w}(k-1)$$
$$\mathbf{h}^{*T}(k) = \mathbf{h}^T(k)T(k)$$

In general, it is common practice to use models with the lowest dimensional order.

A system can be observed when all the states can be determined, directly or indirectly from the measurements. As an algebraic criterion, a sufficient number of linear independent equations is needed. Hence, when a certain state or changes in that state cannot affect the output of the system, the system is said to be unobservable. The observability criterion is satisfied if the matrix M:

$$M = \left(\mathbf{h}\,F^T\mathbf{h}\ldots\left(F^T\right)^{n-1}\mathbf{h}\right) \tag{3.4}$$

has rank n or equivalently the determinant $|M| \neq 0$.

The statistical properties of the system and measurement noise must be defined. It is assumed that they are random quantities, which are not serially correlated and are normally distributed, i.e., they are Gaussian white noise with zero mean and a priori known n^*n system noise covariance matrix $Q(k-1)$ and scalar measurement noise variance $R(k)$:

$$E\{\mathbf{w}(k-1)\} = 0$$

$$E\{\mathbf{w}(k-1)\mathbf{w}^{\mathrm{T}}(l-1)\} = Q(k-1)\delta(k,l)$$

(3.5)

$$E\{v(k)\} = 0$$

$$E\{v(k)v^{\mathrm{T}}(l)\} = R(k)\delta(k,l)$$

$$E\{\mathbf{w}(k-1)v^{\mathrm{T}}(l)\} = 0$$

where $\delta(k,l) = 1$ for $k = l$ and $\delta(k,l) = 0$ for $k \neq l$ is the Kronecker delta and $E\{\}$ denotes the expected value of the quantity between curly brackets for the given probability density function.

3.2 Intermezzo

Any estimate of $\mathbf{x}(k)$, $k = 1,2,\ldots$, which is based on the available measurements $Z(l) = \{z(1),z(2),\ldots,z(l)\}$, is denoted as the n-column vector $\hat{\mathbf{x}}(k/l)$. This notation reads $\hat{\mathbf{x}}(k/l)$ as the estimate of $\mathbf{x}(k)$ at index k given all the measurements up through the index l. Corresponding to the estimate, there is an $n*n$ covariance matrix $P(k/l)$, which involves the n-column estimation error vector $\mathbf{e}(k/l)$ as the difference between the state and its estimate, i.e.:

$$\hat{\mathbf{x}}(k/l) = E\{\mathbf{x}(k)|Z(l)\} = E\{\mathbf{x}(k)\}$$

(3.6)

$$P(k/l) = E\left\{ [\mathbf{x}(k) - \hat{\mathbf{x}}(k/l)][\mathbf{x}(k) - \hat{\mathbf{x}}(k/l)]^{\mathrm{T}} | Z(l) \right\}$$

$$= E\left\{ [\mathbf{x}(k) - \hat{\mathbf{x}}(k/l)][\mathbf{x}(k) - \hat{\mathbf{x}}(k/l)]^{\mathrm{T}} \right\}$$

$$= E\left\{ \mathbf{e}(k/l)\mathbf{e}^{\mathrm{T}}(k/l) \right\}$$

$$\mathbf{e}(k/l) = \mathbf{x}(k) - \hat{\mathbf{x}}(k/l)$$

Depending on the value of k related to l, three types of estimation can be classified:

– **prediction** $(k > l)$, which concerns extrapolation toward future states, i.e., the estimate $\hat{\mathbf{x}}(k/l)$ occurs after the last available measurement.
– **filtering** $(k = l)$, in which only the current state is of interest, i.e., the estimate $\hat{\mathbf{x}}(k/k)$ coincides with the last available measurement.
– **smoothing** $(k < l)$, which involves interpolation of past states, i.e., the estimate $\hat{\mathbf{x}}(k/l)$ falls within a span of available measurements.

As shown, the benefit of the double parenthetical notation is that prediction, filtering, and smoothing can be accommodated within the same notational structure. The price is of course more writing.

3.3   Prediction

Consider the problem of estimating a state of a dynamic system in which the estimated state vector is known at some index $k-1$ with an uncertainty expressed by the covariance matrix:

$$P(k-1/k-1) = E\{\mathbf{e}(k-1/k-1)\mathbf{e}^{\mathrm{T}}(k-1/k-1)\} \tag{3.7}$$

$$= E\{[\mathbf{x}(k-1) - \hat{\mathbf{x}}(k-1/k-1)][\mathbf{x}(k-1) - \hat{\mathbf{x}}(k-1/k-1)]^{\mathrm{T}}\}$$

The estimate $\hat{\mathbf{x}}(k-1/k-1)$ is easily projected ahead by Eq. (3.1) to $\hat{\mathbf{x}}(k/k-1)$ by the transition matrix $F(k,k-1)$. We are justified in ignoring the contribution of the system noise $\mathbf{w}(k-1)$, because it has zero mean and is uncorrelated in the index k. Thus, we have:

$$\hat{\mathbf{x}}(k/k-1) = F(k, k-1)\hat{\mathbf{x}}(k-1/k-1) \tag{3.8}$$

The error vector associated with the prediction $\hat{\mathbf{x}}(k/k-1)$ is obtained by forming:

$$\mathbf{e}(k/k-1) = \mathbf{x}(k) - \hat{\mathbf{x}}(k/k-1) \tag{3.9}$$

$$= F(k, k-1)\mathbf{x}(k-1) + \mathbf{w}(k-1) - F(k, k-1)\hat{\mathbf{x}}(k-1/k-1)$$

$$= F(k, k-1)\mathbf{e}(k-1/k-1) + \mathbf{w}(k-1)$$

It is assumed that the system noise $\mathbf{w}(k-1)$ and estimation error $\mathbf{e}(k-1/k-1)$ are uncorrelated and the system noise covariance matrix $Q(k-1)$ is given by (3.5).

The covariance matrix $P(k/k-1)$ can now be written explicitly as:

$$P(k/k-1) = E\{\mathbf{e}(k/k-1)\mathbf{e}^{\mathrm{T}}(k/k-1)\} \tag{3.10}$$

$$= E\{[F(k, k-1)\mathbf{e}(k-1/k-1) + \mathbf{w}(k-1)][F(k, k-1)\mathbf{e}(k-1/k-1) + \mathbf{w}(k-1)]^{\mathrm{T}}\}$$

$$= F(k, k-1)P(k-1/k-1)F^{\mathrm{T}}(k, k-1) + Q(k-1)$$

In summary, the needed equations for prediction of one-step forward in the sequence from $k-1$ to k are as follows:

$$\hat{\mathbf{x}}(k/k-1) = F(k, k-1)\hat{\mathbf{x}}(k-1/k-1) \tag{3.11}$$

$$P(k/k-1) = F(k, k-1)P(k-1/k-1)F^{\mathrm{T}}(k, k-1) + Q(k-1)$$

To do more steps forward in practice, the following simplified equations are used instead of $\hat{\mathbf{x}}(k+N-1/k-1)$ and $P(k+N-1/k-1)$:

$$\hat{\mathbf{x}}(k/k) = \hat{\mathbf{x}}(k/k-1) \tag{3.12}$$

$$P(k/k) = P(k/k-1)$$

and repeat (3.11) and (3.12) for every step required keeping in mind that for some k no measurements are used. The advantage is that one can account easily for a varying stepsize in prediction.

For retrodiction, the equations for one-step backward from $k+1$ to k are:

$$\hat{\mathbf{x}}_b(k/k+1) = F^{-1}(k+1,k)\hat{\mathbf{x}}_b(k+1/k+1) \tag{3.13}$$

$$P_b(k/k+1) = F^{-1}(k+1,k)\{P_b(k+1/k+1) + Q(k+1)\}F^{-T}(k+1,k)$$

where the superscript "$-T$" means inverse and transpose of a matrix.

To do more steps, backward equations (3.13) are extended for each desired step:

$$\hat{\mathbf{x}}_b(k/k) = \hat{\mathbf{x}}_b(k/k+1) \tag{3.14}$$

$$P_b(k/k) = P_b(k/k+1)$$

The backward retrodiction of the covariance matrix is not as straightforward as the one for the forward prediction. To derive Eq. (3.13), one has to use the continuous differential matrix Riccati equation in its backward form, make the transition to the discrete domain, and solve it accordingly.

3.4 Filtering

Given an a priori estimate of the state $\hat{\mathbf{x}}(k/k-1)$ and covariance matrix $P(k/k-1)$ at the sequence index k. We seek an a posteriori or updated estimate $\hat{\mathbf{x}}(k/k)$ based on the measurement $z(k)$. The state $\mathbf{x}(k)$ is normally distributed and independent of the noise, initialized by its estimate $\hat{\mathbf{x}}(0/0)$ and covariance matrix $P(0/0)$. In order to avoid a growing memory filter, as in classical least squares, the estimate is sought in a recursive form:

new **estimate**$=$*old* **estimate** + **correction**$*$*new* **information**

$$\text{or}: \quad \hat{\mathbf{x}}(k/k) = \mathbf{k}^*(k)\hat{\mathbf{x}}(k/k-1) + \mathbf{k}(k)z(k) \tag{3.15}$$

where $\mathbf{k}^*(k)$ and $\mathbf{k}(k)$ are index-varying weighting matrices as yet unspecified.

An equation for the linear estimator as well as the covariance matrix after incorporation of the new measurement can be obtained from Eq. (3.15) through substitution of the measurement equation (3.2) and the defining estimation error as the difference between the true state $\mathbf{x}(k)$ and its estimate:

$$\mathbf{e}(k/k) = \mathbf{x}(k) - \hat{\mathbf{x}}(k/k) \tag{3.16}$$

$$\mathbf{e}(k/k-1) = \mathbf{x}(k) - \hat{\mathbf{x}}(k/k-1)$$

that after substitution of Eqs. (3.15) and (3.2) take the form:

$$\mathbf{e}(k/k) = \left[I - \mathbf{k}^*(k) - \mathbf{k}(k)\mathbf{h}^T(k)\right]\mathbf{x}(k) + \mathbf{k}^*(k)\mathbf{e}(k/k-1) - \mathbf{k}(k)v(k) \tag{3.17}$$

By definition, $E\{v(k)\}=0$. Also, if $E\{\mathbf{e}(k/k-1)\}=0$, the estimator will be unbiased (i.e., $E\{\mathbf{e}(k/k)\}=0$) for any given state vector $\mathbf{x}(k)$ only if the term between square brackets in Eq. (3.17) is zero. Thus, we demand:

$$\mathbf{k}^*(k) = I - \mathbf{k}(k)\mathbf{h}^T(k) \tag{3.18}$$

and the estimate for the state takes the required form:

$$\hat{\mathbf{x}}(k/k) = \left[I - \mathbf{k}(k)\mathbf{h}^T(k)\right]\hat{\mathbf{x}}(k/k-1) + \mathbf{k}(k)z(k) \tag{3.19}$$

$$\hat{\mathbf{x}}(k/k) = \hat{\mathbf{x}}(k/k-1) + \mathbf{k}(k)\left\{z(k) - \mathbf{h}^T(k)\hat{\mathbf{x}}(k/k-1)\right\}$$

The corresponding estimation error and from here the covariance matrix is given by:

$$\mathbf{e}(k/k) = \left[I - \mathbf{k}(k)\mathbf{h}^T(k)\right]\mathbf{e}(k/k-1) - \mathbf{k}(k)v(k) \tag{3.20}$$

$$P(k/k) = E\left\{[\mathbf{x}(k) - \hat{\mathbf{x}}(k/k)][\mathbf{x}(k) - \hat{\mathbf{x}}(k/k)]^T\right\} = E\left\{\mathbf{e}(k/k)\mathbf{e}^T(k/k)\right\}$$

$$= E\{\left[(I - \mathbf{k}(k)\mathbf{h}^T(k))\mathbf{e}(k/k-1) - \mathbf{k}(k)v(k)\right]$$
$$\cdot\left[(I - \mathbf{k}(k)\mathbf{h}^T(k))\mathbf{e}(k/k-1) - \mathbf{k}(k)v(k)\right]^T\}$$

$$= \left[I - \mathbf{k}(k)\mathbf{h}^T(k)\right]P(k/k-1)\left[I - \mathbf{k}(k)\mathbf{h}^T(k)\right]^T + \mathbf{k}(k)R(k)\mathbf{k}^T(k)$$

where we used the fact that the measurement noise $v(k)$ and estimation error $\mathbf{e}(k/k-1)$ are uncorrelated as well as Eq. (3.5) for the measurement noise variance $R(k)$.

The cost criterion $J(k)$ to minimize $\mathbf{k}(k)$ is chosen as the trace of the covariance matrix. The trace of a square matrix or $tr\{\}$ is the sum of its diagonal elements:

$$J(k) = tr\{P(k/k)\} \tag{3.21}$$

To find the n-column gain vector $\mathbf{k}(k)$, which provides a minimum, it is necessary to take the first partial derivative of the cost criterion $J(k)$ with respect to $\mathbf{k}(k)$, equate it to zero, and solve the resulting equation.

If the dimensions of the matrices X, A, and B match the partial derivatives for the trace of matrix products resulting in a square matrix used here are:

$$\frac{d[tr\{AXB\}]}{dX} = A^T B^T \quad \frac{d[tr\{AX^T B\}]}{dX} = BA \quad \frac{d[tr\{XAX^T\}]}{dX} = X(A + A^T) \tag{3.22}$$

Note that the matrix-vector derivatives by Eq. (2.6) in Section 2.2 are defined in the transposed form.

From Eqs. (3.20)–(3.22) follow:

$$\frac{dJ(k)}{d\mathbf{k}(k)} = -2\left[I - \mathbf{k}(k)\mathbf{h}^{\mathrm{T}}(k)\right]P(k/k-1)\mathbf{h}(k) + 2\mathbf{k}(k)R(k) = 0 \tag{3.23}$$

Solving for the gain vector $\mathbf{k}(k)$ gives immediately:

$$\mathbf{k}(k) = P(k/k-1)\mathbf{h}(k)\left\{\mathbf{h}^{\mathrm{T}}(k)P(k/k-1)\mathbf{h}(k) + R(k)\right\}^{-1} \tag{3.24}$$

which is now referred to as the Kalman gain vector. Although no second partial derivative can be defined, the derived optimal $\mathbf{k}(k)$ is unique and minimizes $J(k)$ for a covariance matrix $P(k/k)$, which is symmetric and positive definite.

From recursive least squares follows that the final equation (3.20) can be transformed for the optimal gain vector $\mathbf{k}(k)$ (3.24) into a somewhat simpler form, which gives for the optimized covariance matrix:

$$P(k/k) = P(k/k-1) - \mathbf{k}(k)\mathbf{h}^{\mathrm{T}}(k)P(k/k-1) \tag{3.25}$$

In summary, for filtering, the estimated state and its covariance matrix are given by:

$$\hat{\mathbf{x}}(k/k) = \hat{\mathbf{x}}(k/k-1) + \mathbf{k}(k)\left\{z(k) - \mathbf{h}^{\mathrm{T}}(k)\hat{\mathbf{x}}(k/k-1)\right\} \tag{3.26}$$

$$P(k/k) = P(k/k-1) - \mathbf{k}(k)\mathbf{h}^{\mathrm{T}}(k)P(k/k-1)$$

$$\mathbf{k}(k) = P(k/k-1)\mathbf{h}(k)\left\{\mathbf{h}^{\mathrm{T}}(k)P(k/k-1)\mathbf{h}(k) + R(k)\right\}^{-1}$$

Note the similarity with the recursive least squares algorithm. They are the same.

For a backward running filter use successively in the filtering equations (3.26), the identities $\hat{\mathbf{x}}_b(k/k) \equiv \hat{\mathbf{x}}(k/k)$, $\hat{\mathbf{x}}_b(k/k+1) \equiv \hat{\mathbf{x}}(k/k-1)$, $P_b(k/k) \equiv P(k/k)$, and $P_b(k/k+1) \equiv P(k/k-1)$.

3.5   Kalman filter

The Kalman filter combines both prediction equation (3.11) and filtering equation (3.26) for the online improvement of the estimated state and its covariance matrix. Note that both equations together are discrete and recursive in nature, i.e., the final results are the starting values for the next computation cycle. It should be clear that once the recursive loop is entered, it could be continued ad infinitum. In practice, however, we always have to deal with a limited number of measurements and the loop ends somewhere.

In summary, the discrete Kalman filter comprises for $k = 1, 2, \ldots$:
Prediction:

$$\hat{\mathbf{x}}(k/k-1) = F(k, k-1)\hat{\mathbf{x}}(k-1/k-1) \tag{3.11}$$

$$P(k/k-1) = F(k, k-1)P(k-1/k-1)F^{\mathrm{T}}(k, k-1) + Q(k-1)$$

Filtering:

$$\hat{\mathbf{x}}(k/k) = \hat{\mathbf{x}}(k/k-1) + \mathbf{k}(k)\{z(k) - \mathbf{h}^{\mathrm{T}}(k)\hat{\mathbf{x}}(k/k-1)\} \qquad (3.26)$$

$$P(k/k) = P(k/k-1) - \mathbf{k}(k)\mathbf{h}^{\mathrm{T}}(k)P(k/k-1)$$

$$\mathbf{k}(k) = P(k/k-1)\mathbf{h}(k)\{\mathbf{h}^{\mathrm{T}}(k)P(k/k-1)\mathbf{h}(k) + R(k)\}^{-1}$$

The state equations of the Kalman filter are also represented in the form of a block diagram as shown in Fig. 3.2.

When no a priori information is available, it is common practice to choose for initialization of the recursion as estimated state the zero vector $\hat{\mathbf{x}}(0/0) = 0$ and as covariance matrix the diagonal matrix $P(0/0) = I \cdot s_0^2$, where I is the $n*n$ identity matrix and the initial variance s_0^2. Compromising statistical bias and numerical stability alike, one has to choose a large value for s_0^2.

A backward running Kalman filter can be expressed in a similar manner.

3.6 Smoothing

State estimation has so far been directed toward prediction and filtering, the state space model, and available measurements are used to estimate the current and future state of the system. Smoothing provides the use of a span of measurements to estimate the history of the state.

For smoothing, a threefold classification has been shown to be useful:

– **fixed-interval smoothing** is a data-processing scheme that uses all the measurements between 1 and N to obtain an estimate $\hat{\mathbf{x}}(k/N)$ with $k < N$.
– **fixed-point smoothing** yields an estimate $\hat{\mathbf{x}}(i/k)$ at a single fixed point $i < k$, which is useful when a previous state of the system is of particular interest.
– **fixed-lag smoothing** gives an estimate $\hat{\mathbf{x}}(k/k+N)$ with a fixed period N backward in the past and the incorporated delay between the current state and the available estimate is inherent to this scheme.

From an algorithmic point of view, the various types of smoothing are quite similar.

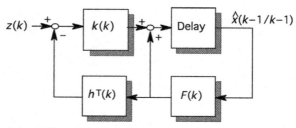

FIG. 3.2 The Kalman filter.

3.6.1   Fixed-interval smoothing

Fixed-interval smoothing is a data-processing scheme to obtain an estimate $\hat{\mathbf{x}}(k/N)$ backward in the past with $k < N$. Here, smoothing can only be done off-line afterward, because it must be performed after all the N measurements have been collected. This type of smoothing can be considered as the combination of two Kalman filters. Firstly, a forward Kalman filter operates on the measurements, for which the filtered state $\hat{\mathbf{x}}(k/k)$ and its covariance matrix $P(k/k)$ are stored at each step. In a similar way, a second Kalman filter operates backward in the sequence with the retrodiction $\hat{\mathbf{x}}_b(k/k+1)$ and $P_b(k/k+1)$, respectively. Thereafter, the smoothed estimate follows from the statistical weighting or pooling of the stored estimates by means of their stored covariance matrices. The smoothed estimates $\hat{\mathbf{x}}(k/N)$ and covariance matrices $P(k/N)$ can be calculated for $k = 1,2,...,N-1$ by:

$$\hat{\mathbf{x}}(k/N) = P(k/N)\left[P^{-1}(k/k)\hat{\mathbf{x}}(k/k) + P_b^{-1}(k/k+1)\hat{\mathbf{x}}_b(k/k+1)\right] \quad (3.27)$$

$$P(k/N) = \left[P^{-1}(k/k) + P_b^{-1}(k/k+1)\right]^{-1}$$

For the smoothed state, all the measurements between 1 and N are taken into account because the estimate based on the measurements up to a given point k is combined with the estimate based upon the measurements from the final N backward to the point $k+1$ and subsequently retrodicted to point k. Eq. (3.27) indicates considerable additional requirements of memory storage and computation time beyond that are required for the Kalman filter. The algorithm can be shown to be equivalent to some alternative formulations in order to reduce the computational efforts of the matrix inversions and memory storage.

The simplest and most elegant form of fixed-interval smoothing is the one after Rauch, Tung, and Striebel (RTS-smoother). The forward run is provided by a conventional Kalman filter processing the measurements online. Hereafter, the backward Kalman filter and pooling of the estimates are modified to process the stored quantities in a reverse sweep for $k = N-1, N-2,...,1$:

$$\hat{\mathbf{x}}(k/N) = \hat{\mathbf{x}}(k/k) + A(k)[\hat{\mathbf{x}}(k+1/N) - \hat{\mathbf{x}}(k+1/k)] \quad (3.28)$$

$$P(k/N) = P(k/k) + A(k)[P(k+1/N) - P(k+1/k)]A^T(k)$$

$$A(k) = P(k/k)F^T(k+1,k)P^{-1}(k+1/k)$$

where for $k = N$: $\hat{\mathbf{x}}(k/N) = \hat{\mathbf{x}}(k/k)$ and $P(k/N) = P(k/k)$. $A(k)$ refers to the $n*n$ smoother gain matrix. The smoother is initialized by means of the last filtered state $\hat{\mathbf{x}}(k/k)$ and covariance matrix $P(k/k)$ of the Kalman filter. The state equation of fixed-interval smoothing is depicted as a block diagram in Fig. 3.3.

The computational effort is now limited to one matrix inversion at each step.

The need for memory storage can be limited to the state estimate $\hat{\mathbf{x}}(k/k)$ and covariance matrix $P(k/k)$, because $\hat{\mathbf{x}}(k+1/k)$ and $P(k+1/k)$ can be computed by prediction. By using the symmetry of the off-diagonal elements in the covariance matrix $P(k/k)$, the memory requirements can be reduced further. A disadvantage

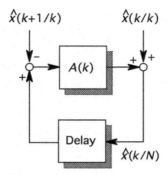

FIG. 3.3 Fixed-interval smoothing.

here still lies in the matrix inversion at each step and numerical problems may arise when $P(k+1/k)$ becomes ill-conditioned.

A state is said to be smoothable if the optimal smoother provides a state estimate superior to that obtained when the final optimal filter estimate is extrapolated backward. Eq. (3.27) indicates the restriction $P(k/N) \leq P(k/k)$. Only states that are driven by system noise are smoothable. If the system noise covariance matrix $Q(k)$ equals the zero matrix, then the smoother gain matrix becomes simply:

$$A(k) = F(k, k+1) = F^{-1}(k+1, k) \tag{3.29}$$

$$\text{and}: \quad \hat{x}(k/N) = F^{-1}(k+1, k)\hat{x}(k+1/N) \tag{3.30}$$

$$P(k/N) = F^{-1}(k+1, k)P(k+1/N)F^{-T}(k+1, k)$$

If in addition also $F^{-1}(k+1, k) = I$, then the estimate $\hat{x}(k/N)$ and covariance matrix $P(k/N)$ remain constant in the sequence running backward.

3.6.2 Fixed-point smoothing

Fixed-point smoothing yields an estimate $\hat{x}(i/k)$ at a single fixed point $i < k$, which is useful when a previous state of the system at a fixed point is of particular interest. We present just one algorithm because of its simplicity and similarity to fixed-interval smoothing. Starting from the initial condition at $k = i$: $\hat{x}(i/k) = \hat{x}(k/k)$ and $P(i/k) = P(k/k)$, the estimate $\hat{x}(i/k)$ and its covariance matrix $P(i/k)$ are processed from quantities provided by a concurrently forward running Kalman filter. With definition of the $n*n$ gain matrix $B(k,i)$, we have the following equations:

$$\hat{x}(i/k) = \hat{x}(i/k-1) + B(k, i)[\hat{x}(k/k) - \hat{x}(k/k-1)] \tag{3.31}$$

$$P(i/k) = P(i/k-1) + B(k, i)[P(k/k) - P(k/k-1)]B^T(k, i)$$

$$B(k,i) = \prod_{j=i}^{k-1} A(j) = B(k-1,i)A(k-1)$$

$$A(j) = P(j/j)F^{T}(j+1,j)P^{-1}(j+1/j)$$

$$i \text{ fixed and } k = i+1, i+2, \ldots$$

A diagram for the estimated state of fixed-point smoothing is shown in Fig. 3.4.

The difference between the estimates $\hat{\mathbf{x}}(k/k) - \hat{\mathbf{x}}(k/k-1)$ as well as the covariance matrices $P(k/k) - P(k/k-1)$ is obtained directly from the Kalman filter (3.26).

This gives finally for fixed-point smoothing the equations:

$$\hat{\mathbf{x}}(i/k) = \hat{\mathbf{x}}(i/k-1) + B(k,i)\mathbf{k}(k)\{z(k) - \mathbf{h}^{T}(k)\hat{\mathbf{x}}(k/k-1)\} \qquad (3.32)$$

$$P(i/k) = P(i/k-1) - B(k,i)\mathbf{k}(k)\mathbf{h}^{T}(k)P(k/k-1)B^{T}(k,i)$$

To avoid the inversion of the covariance matrix $P(k/k-1)$ as required for each recursion alternative algorithms can be used, which are not presented here.

3.6.3   Fixed-lag smoothing

An estimate $\hat{\mathbf{x}}(k/k+N)$ following the state with a fixed period N backward in the past is governed by fixed-lag smoothing. The incorporated delay between the current state and the available estimate is inherent to this scheme. An algorithm for fixed-lag smoothing is somewhat more complicated than the ones for the two other smoothing categories. However, knowing the RTS-algorithm for fixed-interval smoothing (3.28), one can always do fixed-lag smoothing by first filtering online the current measurement and then sweeping a fixed number of steps N backward in the past. In such way, one generates the quantities $\hat{\mathbf{x}}(k/k+N)$ and $P(k/k+N)$. If the number of steps backward is small and the

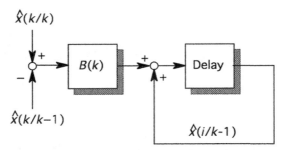

FIG. 3.4   Fixed-point smoothing.

dimension of the model is not too high, this is a simple and effective way of doing fixed-lag smoothing; otherwise, this method can become time consuming and one should look for more efficient algorithms.

It should be noted that once the described fixed-lag smoothing is in operation the number of matrix inversions can be limited to one each step forward, because the previous gain matrices can be stored and reused if they are needed again.

3.7 Examples

Example 3.1: The mathematical interpretation of a slowly varying state indicates a time-dependent disturbance. Linear drift can often be used. A single state $x(k)$ affected by linear drift can be described by:

$$x(k) = dk + e \tag{3.33}$$

where d is the drift parameter and e is the value of $x(k)$ at $k = 0$.

The state space model is based on the static description:

$$\begin{pmatrix} e(k) \\ d(k) \end{pmatrix} = \begin{pmatrix} 1 & 0 \\ 0 & 1 \end{pmatrix} \begin{pmatrix} e(k-1) \\ d(k-1) \end{pmatrix} + \mathbf{w}(k-1) \tag{3.34}$$

$$z(k) = (1 \quad k) \begin{pmatrix} e(k) \\ d(k) \end{pmatrix} + v(k)$$

A disadvantage of using this model is that the desired state depends on the index k. Consequently, an alternative representation has to be developed. Reformulating (3.33) and introduction of system and measurement noise give the following state space model:

$$x(k) = d \cdot (k-1) + d + e = x(k-1) + d \tag{3.35}$$

$$\begin{pmatrix} x(k) \\ d(k) \end{pmatrix} = \begin{pmatrix} 1 & 1 \\ 0 & 1 \end{pmatrix} \begin{pmatrix} x(k-1) \\ d(k-1) \end{pmatrix} + \mathbf{w}(k-1) \tag{3.36}$$

$$z(k) = (1 \quad 0) \begin{pmatrix} x(k) \\ d(k) \end{pmatrix} + v(k)$$

Important features of this model are that it incorporates a dynamic structure and involves the required parameters directly in the state. Both state space models (3.34) and (3.36) are related by the similarity matrix:

$$T(k) = \begin{pmatrix} 1 & -k \\ 0 & 1 \end{pmatrix} \tag{3.37}$$

A drifting system can thus be described by using two random state variables. The alternative model is called stochastic drift, represented by the combination of a random constant, random ramp, and random walk. In Fig. 3.5, an example

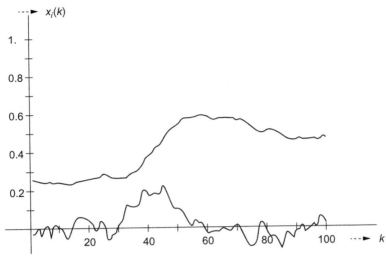

FIG. 3.5 An example of a system involving stochastic drift. Upper curve $x()$ and lower curve $10*d()$.

of a system involving stochastic drift is given. For plotting reasons, the strongly varying state variable $d(k)$ around zero is multiplied by a factor 10. The measurements including stochastic drift and measurement noise are shown in Fig. 3.6. For the system noise variances, the values $Q_{11}(k-1)=2\cdot10^{-5}$ and $Q_{22}(k-1)=8\cdot10^{-6}$ are employed. The measurement noise variance $R(k)=10^{-5}$ holds. The estimated state by the Kalman filter, fixed-interval smoothing, fixed-point smoothing for $i=1$ and fixed-lag smoothing with a delay $N=10$ follow in Figs. 3.7–3.10.

Example 3.2: A periodic or oscillatory system can be described by the differential equation:

$$\ddot{x}+\omega^2x=0 \tag{3.38}$$

where x denotes the response at time t, $\ddot{x}$ the second derivative, and ω is the periodic constant. Let us assume that the measurements are taken discrete and uniformly spaced with a sampling time Δt. The state space model then becomes:

$$\begin{pmatrix}x_1(k)\\x_2(k)\end{pmatrix}=\begin{pmatrix}\cos\omega\Delta t & \sin\omega\Delta t\\-\sin\omega\Delta t & \cos\omega\Delta t\end{pmatrix}\begin{pmatrix}x_1(k-1)\\x_2(k-1)\end{pmatrix}+\mathbf{w}(k-1) \tag{3.39}$$

$$z(k)=(1\ \ 0)\begin{pmatrix}x_1(k)\\x_2(k)\end{pmatrix}+v(k)$$

A periodic system can thus be described by using two random state variables.

A disadvantage of this state space model is that the periodic term $\omega\Delta t$ must be known in advance for use within the transition matrix. In Fig. 3.11, the

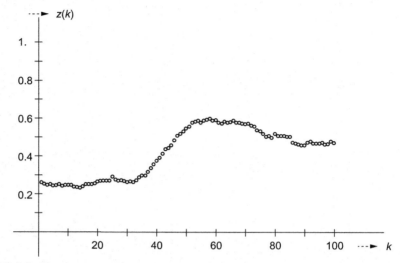

FIG. 3.6   The measurements.

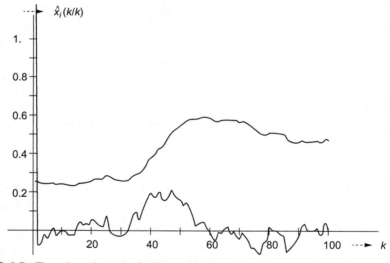

FIG. 3.7   The estimated state by the Kalman filter (see Fig. 3.5).

measurements are plotted for the term $\omega\Delta t = 0.3142$, a measurement noise variance $R(k) = 10^{-4}$, and system noise variances $Q_{11}(k-1) = Q_{22}(k-1) = 10^{-4}$. In Fig. 3.12, the estimated state by exclusively the Kalman filter is depicted. In Chapter 5, this model extended with the periodic term $\omega\Delta t$ serves as a depicted nonlinear example.

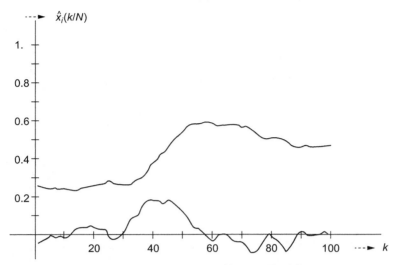

FIG. 3.8 The estimated state by fixed-interval smoothing (see Fig. 3.5).

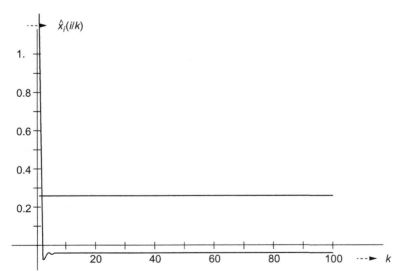

FIG. 3.9 The estimated state by fixed-point smoothing for $i = 1$ (see Fig. 3.5).

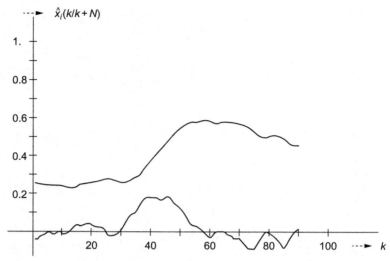

FIG. 3.10   The estimated state by fixed-lag smoothing with a delay $N = 10$ (see Fig. 3.5).

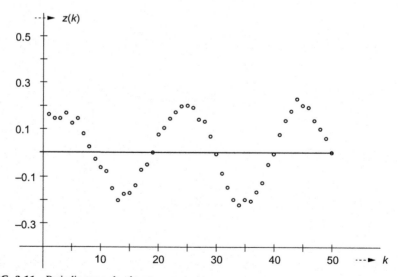

FIG. 3.11   Periodic example: the measurements.

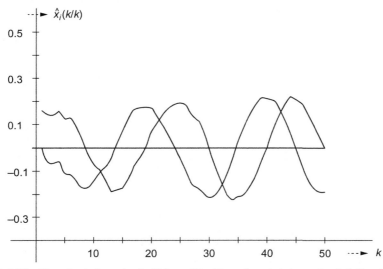

FIG. 3.12   The estimated state by the Kalman filter. From above to below at $k = 1$: $\hat{x}_1()$ and $\hat{x}_2()$.

Example 3.3: An exponential system or first-order process can be described by the differential equation:

$$\dot{x} + 1/T_x x = x_\infty \tag{3.40}$$

where x describes the response at time t, $\dot{x}$ the first derivative, T_x the time constant, and x_∞ the value at infinity. The continuous solution of the differential equation is:

$$x(t) = x_\infty + (x(0) - x_\infty)e^{-t/T_x} \tag{3.41}$$

For a sampling time Δt, introducing the exponential factor $\Phi = e^{-\Delta t/T_x}$, the system can be represented in discrete time terms by:

$$x(t + \Delta t) = \Phi x(t) + (1 - \Phi)x_\infty \tag{3.42}$$

The state space model thus becomes:

$$\begin{pmatrix} x(k) \\ x_\infty(k) \end{pmatrix} = \begin{pmatrix} \Phi & 1-\Phi \\ 0 & 1 \end{pmatrix} \begin{pmatrix} x(k-1) \\ x_\infty(k-1) \end{pmatrix} + \mathbf{w}(k-1) \tag{3.43}$$

$$z(k) = (1 \quad 0) \begin{pmatrix} x(k) \\ x_\infty(k) \end{pmatrix} + v(k)$$

An exponential system can thus be described by using two random state variables.

A disadvantage of this model is that Φ must be known in advance for use within the transition matrix. In Fig. 3.13, the measurements are plotted for an exponential factor $\Phi = 0.9$, infinite value $x_\infty = 0.5$, measurement noise variance $R(k) = 10^{-5}$, and system noise variances $Q_{11}(k-1) = Q_{22}(k-1) = 10^{-6}$. In Fig. 3.14, the estimated state by exclusively the Kalman filter is depicted. In Chapter 5, this model is extended to cover the exponential factor Φ in a nonlinear state space model.

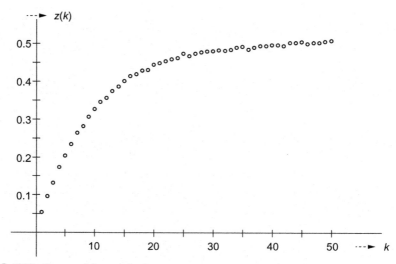

FIG. 3.13   Exponential example: the measurements.

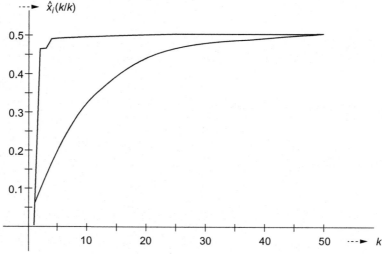

FIG. 3.14   The estimated state by the Kalman filter. From above to below at $k = 20$: $\hat{x}_\infty()$ and $\hat{x}()$.

Example 3.4: As a last example, consider for a kinetic system the first-order consecutive reaction $A \rightarrow B \rightarrow C$ with k_1 and k_2 as the reaction constants. For this kinetic system, the following coupled differential equations can be formed:

$$\dot{x}_1 = -k_1 x_1 \tag{3.44}$$

$$\dot{x}_2 = k_1 x_1 - k_2 x_2$$

$$\dot{x}_3 = k_2 x_2$$

where x_i are the concentrations of component i at time t; [A], [B], and [C] and $\dot{x}_i$ are the first derivatives.

For a sampling time Δt, the solution of this differential system is written down directly in discrete time terms:

$$F(k, k-1) = \begin{pmatrix} e^{-k_1 \Delta t} & 0 & 0 \\ \dfrac{k_1}{k_2 - k_1}\{e^{-k_1\Delta t} - e^{-k_2\Delta t}\} & e^{-k_2\Delta t} & 0 \\ 1 + \dfrac{k_1}{k_2 - k_1}e^{-k_2\Delta t} - \dfrac{k_2}{k_2 - k_1}e^{-k_1\Delta t} & 1 - e^{-k_2\Delta t} & 1 \end{pmatrix} \tag{3.45}$$

where $\mathbf{x}(k) = F(k, k-1)\mathbf{x}(k-1) + \mathbf{w}(k-1)$ and $\mathbf{x}^T(k) = (x_1(k) \; x_2(k) \; x_3(k))$.

If the kinetic reaction is followed spectroscopically, the measurement vector reads: $\mathbf{h}^T(k) = (\varepsilon_1 \; \varepsilon_2 \; \varepsilon_3)$ with ε_i as the molar absorptivities of component i.

The measurement equation holds: $z(k) = \mathbf{h}^T(k)\mathbf{x}(k) + v(k)$. It should be noted that the reaction constants k_1 and k_2 as well as the molar absorptivities ε_1, ε_2, and ε_3 have to be known in advance for use in the state space model. The system is observable if the reaction is followed at one wavelength with the necessity $\varepsilon_3 \neq 0$. The contributions of ε_1 and ε_2 are not needed to follow the kinetic reaction.

System noise variances are $Q_{11}(k-1) = Q_{22}(k-1) = Q_{33}(k-1) = 10^{-6}$ and for the measurement noise variance $R(k) = 10^{-5}$ is used. For convenience, $\mathbf{h}^T(k) = (0 \; 0 \; 1)$ is assumed. As example $k_1 = 0.05$, $k_2 = 0.1$, and $\Delta t = 1$ are employed resulting in a transition matrix:

$$F(k, k-1) = \begin{pmatrix} 0.9512 & 0 & 0 \\ 0.0464 & 0.9048 & 0 \\ 0.0024 & 0.0952 & 1 \end{pmatrix} \tag{3.46}$$

In Fig. 3.15 the simulated states for a first-order consecutive reaction are given, and in Fig. 3.16 the resulting measurements with noise are shown. In Figs. 3.17 and 3.18, the estimated state by the Kalman filter and fixed-interval smoothing are plotted.

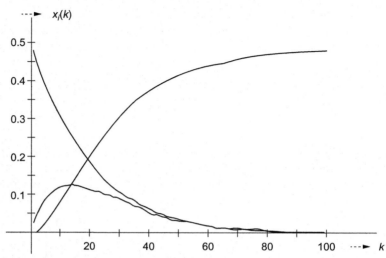

FIG. 3.15   An example of a system involving consecutive kinetics. At $k = 10$ from above to below: $x_1()$, $x_2()$, and $x_3()$.

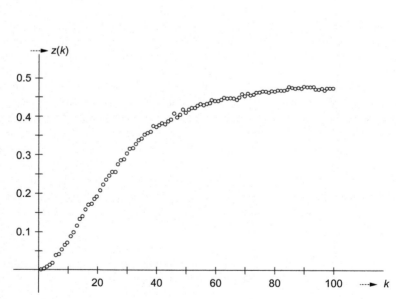

FIG. 3.16   The measurements.

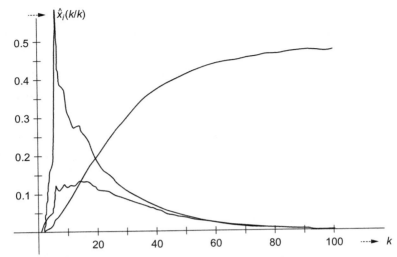

FIG. 3.17   The estimated state by the Kalman filter (see Fig. 3.15).

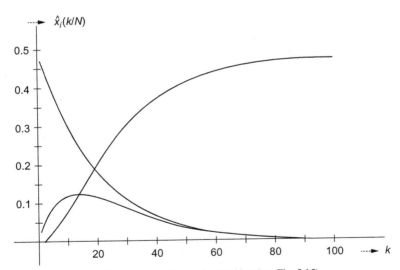

FIG. 3.18   The estimated state by fixed-interval smoothing (see Fig. 3.15).

Chapter 4

Statistics

Chapter outline

4.1 Verification

Within state estimation, the statistics for the verification of the results are rather simple and straightforward. The innovation $v(k)$ is defined as the difference between the experimental and estimated measurement:

$$v(k) = z(k) - \mathbf{h}^{\mathrm{T}}(k)\hat{\mathbf{x}}(k/k-1) \tag{4.1}$$

The innovation sequence has been shown to be Gaussian white noise:

$$E\{v(k)\} = 0 \tag{4.2}$$

$$E\{v(k)v^{\mathrm{T}}(l)\} = \{\mathbf{h}^{\mathrm{T}}(k)P(k/k-1)\mathbf{h}(k) + R(k)\}\delta(k, l)$$

The statistical characteristics of the monitored innovations can be online exploited in practice for failure detection, checking for a correct model implemented or for identifying unknown noise variances. For testing purposes, it is more convenient to consider the standardized innovations, which are normally distributed with mean 0 and variance 1:

$$n(k) = v(k)\{\mathbf{h}^{\mathrm{T}}(k)P(k/k-1)\mathbf{h}(k) + R(k)\}^{-1/2} \tag{4.3}$$

The mean $\bar{n}(k)$ of the standardized innovations is given by:

$$\bar{n}(k) = \frac{1}{k}\sum_{i=1}^{k} n(i) = (1 - 1/k)\bar{n}(k-1) + 1/k \cdot n(k) \tag{4.4}$$

where k denotes the sample size. Given the Gaussian whiteness of the standardized innovations, $\bar{n}(k)$ is normally distributed with zero mean and variance $1/k$. This result can be used immediately for online verification.

State Estimation in Chemometrics. https://doi.org/10.1016/B978-0-08-102603-8.00004-3

The sum of the squared standardized innovations or the quantity:

$$\chi^2(k-n) = \sum_{i=1}^{k} n(i)^2 = \chi^2(k-n-1) + n(k)^2 \qquad (4.5)$$

has a chi-square distribution with $k-n$ degrees of freedom. This result can be used directly for online verification, where the computed $\chi^2(k-n)$-values are compared with their critical limits. Chi-square statistics are employed as a two-sided test for a given confidence level α and $k-n$ degrees of freedom. Modeling errors may cause excessive χ^2-values that also may result from the use of insufficient noise variances. Alternatively, too low χ^2-values may be associated with the implementation of higher-valued noise variances as required. Critical table values of chi-square can be found in statistical handbooks.

As a criterion for the optimal nature of the Kalman filter, the autocorrelogram $\phi(i)$, $i = 1,2,\dots$ of the N standardized innovations can be computed offline after all the measurements have been processed. A prerequisite is that the measurements are equidistant in the sequence dependent variable:

$$\phi(i) = \frac{1}{N-i-1} \sum_{k=1}^{N-i} (n(k) - \bar{n})(n(k+i) - \bar{n})/s^2 \qquad (4.6)$$

where the variance is: $s^2 = \frac{1}{N-1}\sum_{k=1}^{N}(n(k) - \bar{n})^2$ and the mean $\bar{n} = \frac{1}{N}\sum_{k=1}^{N}n(k)$.

For Gaussian white standardized innovations, it can be shown that the computed $\phi(i)$, $i = 1,2,\dots$ are normally distributed with zero mean and variance $1/(N-i)$.

For the statistical results from fixed-interval smoothing, the residuals $r(k)$ can be used in the reverse way starting at $k = N$ and going backward to $k = 1$:

$$r(k) = z(k) - \mathbf{h}^T(k)\hat{\mathbf{x}}(k/N) \qquad (4.7)$$

For the smoother residuals, Eq. (4.3) with the identities $r(k) \equiv v(k)$ and $P(k/N) \equiv P(k/k-1)$ is further employed; the rest of the verification stays the same.

4.2 Evaluation

A linear projection in multidimensional space of the estimated state vector and the covariance matrix can be written as:

$$\hat{\mathbf{x}}^*(k/k) = C\hat{\mathbf{x}}(k/k) \qquad (4.8)$$

$$P^*(k/k) = CP(k/k)C^T$$

where r is the rank of the p^*n projection matrix C with $1 \leq p \leq n$.

The confidence region of the estimate $\hat{\mathbf{x}}^*(k/k)$ is given by:

$$(\mathbf{x}(k) - \hat{\mathbf{x}}^*(k/k))^T P^{*-1}(k/k)(\mathbf{x}(k) - \hat{\mathbf{x}}^*(k/k)) = rF(1-\alpha, r, k-n) \qquad (4.9)$$

where $F(1-\alpha, r, k-n)$ is the Fisher F-value for a given confidence level α and $(r, k-n)$ degrees of freedom. Critical values can be found in statistical handbooks.

For smoothing use in (4.8)–(4.10) and (4.12), the identities $\hat{\mathbf{x}}(k/N) \equiv \hat{\mathbf{x}}(k/k)$, $\hat{\mathbf{x}}^*(k/N) \equiv \hat{\mathbf{x}}^*(k/k)$, $P(k/N) \equiv P(k/k)$, and $P^*(k/N) \equiv P^*(k/k)$. For C, the similarity matrix $T^{-1}(k)$ can be chosen. If C equals the $n*n$ identity matrix I, the confidence region is obtained for the whole state vector in multidimensional space. The confidence interval for one element $x_i(k)$ of the state vector $\mathbf{x}(k)$ follows by setting C to the $n*n$ null matrix with one diagonal element C_{ii} equal to 1 making the rank of the matrix $r = 1$. Because $F(1-\alpha, 1, k-n) = t^2(1-\alpha/2, k-n)$, this gives as a confidence interval for the element $x_i(k)$ of the state $\mathbf{x}(k)$ to be used for evaluation:

$$\hat{x}_i(k/k) \pm t(1-\alpha/2, k-n)\sqrt{P_{ii}(k/k)} \qquad (4.10)$$

Also a Student's t-test can be used for verification if an estimated element differs significantly from a known true value. For state estimation implemented with a priori known noise variances $R(k)$ and $Q(k-1)$, the critical Fisher F-value and Student's t value become $F(1-\alpha, r, \infty)$ and $z(1-\alpha/2) = t(1-\alpha/2, \infty)$, respectively.

Here, $z(1-\alpha/2)$ is the critical z-value for a given confidence level α.

4.3 Selection

Based on a given confidence level α, an F-test can check one-sided whether two independent χ^2-values each with its degrees of freedom are statistically different:

$$F(k_1, k_2) = \chi_1^2(k_1)/k_1 \cdot k_2/\chi_2^2(k_2) \qquad (4.11)$$

where k_1 and k_2 are the possible different number of degrees of freedom.

In this way, it is possible to compare different models, estimation techniques, or data preprocessing procedures and select the right one.

An F-test verifies the relation $\mathbf{a} = C\hat{\mathbf{x}}(k/k)$, which is tested by:

$$(\mathbf{a} - \hat{\mathbf{x}}^*(k/k))^T P^{*-1}(k/k)(\mathbf{a} - \hat{\mathbf{x}}^*(k/k)) \leq rF(1-\alpha, r, k-n) \qquad (4.12)$$

In first instance, verification can be used for checking the estimated state with true values known from simulation or reference values. Here, C is the $n*n$ null matrix with $C_{ii} = 1$ for each verified element i and a_i as the tested true value. It should be noted that this test could also be used for selection; in case if there are elements in the estimated state not significantly different from zero. Here again, C is the $n*n$ null matrix with for each element i tested $C_{ii} = 1$ and $a_i = 0$ as the testing value.

For a priori known noise variances, the critical F-value reduces to $F(1-\alpha, r, \infty)$.

4.4 Normality

The standardized innovations or residuals can be tested offline afterward against a normal distribution. There are a lot of tests for normality; here three are employed: the skewness-kurtosis test, chi-square test, and Kolmogorov-Smirnov test.

4.4.1 Skewness-kurtosis test

For the standardized innovations or residuals, the skewness g_1 and kurtosis g_2 are defined as:

$$g_1 = \frac{1}{N}\sum_{k=1}^{N}\left(\frac{n(k) - \bar{n}}{s}\right)^3 \tag{4.13}$$

$$g_2 = \frac{1}{N}\sum_{k=1}^{N}\left(\frac{n(k) - \bar{n}}{s}\right)^4 - 3$$

For a reasonable large number of Gaussian white data, both the skewness and kurtosis are normally distributed with a zero mean and given variance:

$$\sigma^2(g_1) = \frac{6N(N-1)}{(N-2)(N+1)(N+3)} \tag{4.14}$$

$$\sigma^2(g_2) = \frac{24N(N-1)^2}{(N-2)(N-3)(N+3)(N+5)}$$

For a normality test, one can either use the z-values $z_1 = g_1/\sigma(g_1)$ and $z_2 = g_2/\sigma(g_2)$ or combine these values in a chi-square test given by $\chi^2(2) = z_1^2 + z_2^2$, which is tested one-sided against a critical table value with two degrees of freedom. Table values with the critical limits for a given confidence level α of the different tests can be found in statistical handbooks.

4.4.2 Chi-square test

This goodness-of-fit test is based upon frequencies of observations divided over ℓ classes in a histogram, which are compared with an expected theoretical distribution:

$$\chi^2(\ell - p) = \sum_{i=1}^{\ell}\frac{(O_i - E_i)^2}{E_i} \tag{4.15}$$

with: $O_i = $ observed number of observations in class i.
$E_i = $ expected number of observations in class i (minimal 5).
Here, p denotes the number of parameters involved for the determination of the expected theoretical distribution. The computed chi-square value by Eq. (4.15) is tested one-sided against a critical statistical table value for a given

confidence level α and $\ell - p$ degrees of freedom. For the determination of a normal frequency distribution, three parameters $\bar{n}$, s^2, and N are involved, i.e., $p = 3$.

4.4.3 Kolmogorov-Smirnov test

This test is based upon a cumulative frequency distribution. The observations $n(i)$ are ordered according to increasing values given by $x(i)$, $i = 1, 2, ..., N$. For the observations $x(i)$, the cumulative frequency function $S_N(x)$ is defined as:

$$S_N(x) = i/N \quad \text{for } x(i) \le x \le x(i+1) \tag{4.16}$$

Each observation $x(i)$ is standardized by using the mean and variance according to $(x(i) - \bar{n})/s$, and from here the cumulative normal probability $P\{x(i)\}$ is computed. The maximal absolute difference is compared one-sided with a critical statistical table value for a given confidence level α and number of observations N:

$$D_{\max} = \max\left[|\, P\{x(i)\} - S_N(x(i))|\,,\, |\, P\{x(i)\} - S_N(x(i-1))|\right] \ i = 1, 2, ..., N \tag{4.17}$$

where "| |" denotes the absolute value. The Kolmogorov-Smirnov test demands fewer observations in comparison with the chi-square test.

4.5 Example

Example 4.1: Statistical results are exclusively given for the stochastic drift system operating with the Kalman filter in Example 3.1. In Fig. 4.1, the standardized innovations resulting from the Kalman filter are plotted, which are transformed to the histogram in Fig. 4.2. In addition, the autocorrelogram and its 95% confidence limits are depicted in Fig. 4.3.

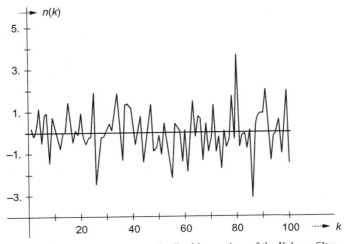

FIG. 4.1 Stochastic drift example: the standardized innovations of the Kalman filter.

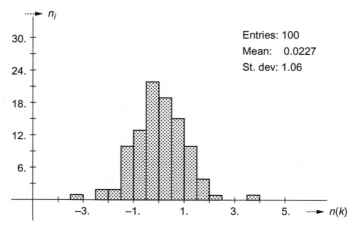

FIG. 4.2   Histogram of the standardized innovations.

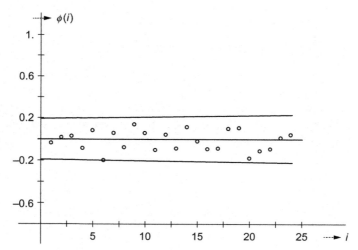

FIG. 4.3   Autocorrelogram of the standardized innovations and its 95% confidence limits.

Normality testing is done from Figs. 4.4 to 4.7. In Fig. 4.4, the Kolmogorov-Smirnov test is performed. The cumulative frequency function of the observed values is compared with the cumulative normal distribution in order to determine the maximal difference. In Figs. 4.5 and 4.6, the histograms with the

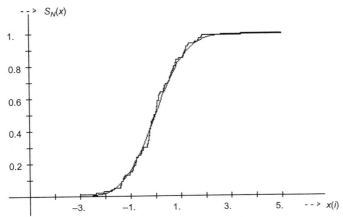

FIG. 4.4 The Kolmogorov-Smirnov test: the observed cumulative frequency function in comparison with the cumulative normal distribution.

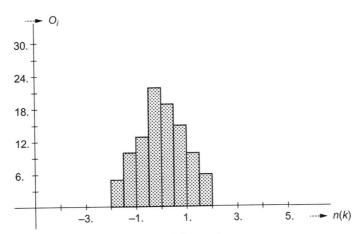

FIG. 4.5 Chi-square test: histogram observed frequencies.

observed frequencies and expected frequencies for a normal distribution have to be compared for the chi-square test. An example of a printout for normality testing is given in Fig. 4.7. It shows for a 95% confidence level the normality of the standardized innovations (with exception of the kurtosis).

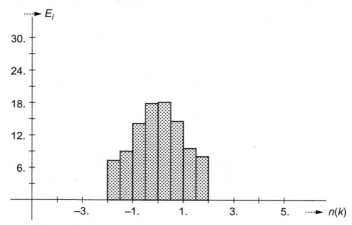

FIG. 4.6 Chi-square test: histogram expected frequencies for a normal distribution.

Normality
Skewness and kurtosis test
Entries: 100 Mean: 0.0227 St. dev: 1.06
Skewness 0.0628 St. dev. 0.241 Z-value 0.26
Kurtosis 1.11 St. dev. 0.478 Z-value 2.33
Combined chi-square 5.48
Critical X2 (95%, 2) = 5.99

Kolmogorov-Smirnov test
Cumulative normal distribution
Entries   100 Mean   0.0227 St. dev. 1.06
Greatest difference   0.0735
Critical D(5%, 100) = 0.136

Histogram chi-square test
for normal distribution
Entries: 100 Mean: 0.0227 St. dev: 1.06
Chi-square value 2.49
Critical X2 (95%, 5) = 11.1

FIG. 4.7 Printout for the applied normality tests.

In Figs. 4.8 and 4.9, the computed online means and chi-square values of the standardized innovations with their 95% confidence limits are given. From here, it follows that the implemented state space model and noise variances explain the standardized innovations in a statistically correct way.

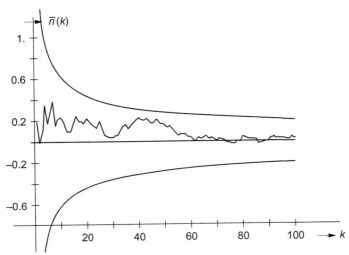

FIG. 4.8 The online means of the standardized innovations with their 95% confidence limits.

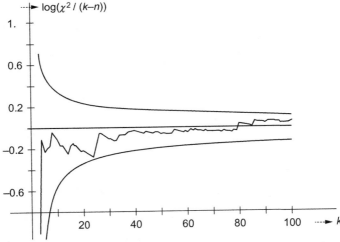

FIG. 4.9 The online chi-square values of the standardized innovations with their 95% confidence limits.

All these results together demonstrate clearly for the stochastic drift system the Gaussian whiteness of the standardized innovations coming from the Kalman filter.

Similar results can be generated for every other example in this book.

Chapter 5

Nonlinear estimation

Chapter outline

5.1 Modeling

There are many instances where the system equation and/or the measurement equation cannot be expressed in a simple linear form as required by the state space model. A more general form for these nonlinear models is given by:

$$\mathbf{x}(k) = f(\mathbf{x}(k-1), k-1) + \mathbf{w}(k-1) \tag{5.1}$$

$$z(k) = g(\mathbf{x}(k), k) + v(k)$$

where $f()$ and $g()$ are nonlinear functions of the state.

5.2 Extended Kalman filter

The derivation of the equations for nonlinear estimation is rather tedious and cumbersome and shall not be given. We just focus here on the resulting equations.

The equations for $f()$ and $g()$ can be cast into an approximate linear form by expanding them into a Taylor series, which are truncated after the second linear term:

$$f(\mathbf{x}(k-1), k-1) = f(\hat{\mathbf{x}}(k-1/k-1), k-1) + F(\hat{\mathbf{x}}(k-1/k-1), k, k-1)$$
$$\cdot \{\mathbf{x}(k-1) - \hat{\mathbf{x}}(k-1/k-1)\} + h.o.t. \tag{5.2}$$

$$g(\mathbf{x}(k), k) = g(\hat{\mathbf{x}}(k/k-1), k) + \mathbf{h}^{\mathrm{T}}(\hat{\mathbf{x}}(k/k-1), k)\{\mathbf{x}(k) - \hat{\mathbf{x}}(k/k-1)\} + h.o.t.$$

The partial derivatives, i.e., the transition matrix $F()$ and transposed measurement vector $\mathbf{h}^{\mathrm{T}}()$ are defined by:

$$F(\hat{\mathbf{x}}(k-1/k-1), k, k-1) = \left.\frac{df(\mathbf{x}(k-1), k-1)}{d\mathbf{x}(k-1)}\right|_{\mathbf{x}(k-1)=\hat{\mathbf{x}}(k-1/k-1)} \tag{5.3}$$

State Estimation in Chemometrics. https://doi.org/10.1016/B978-0-08-102603-8.00005-5

$$\mathbf{h}^{\mathrm{T}}(\hat{\mathbf{x}}(k/k-1), k) = \frac{dg(\mathbf{x}(k), k)}{d\mathbf{x}(k)}\Bigg|_{\mathbf{x}(k)=\hat{\mathbf{x}}(k/k-1)}$$

These partial derivatives are evaluated for the most recent values of the estimated state. The estimation equations are now implemented in the same way as for the linear model with some minor modifications.

This results in the extended Kalman filter:

Prediction:

$$\hat{\mathbf{x}}(k/k-1) = f(\hat{\mathbf{x}}(k-1/k-1), k-1) \tag{5.4}$$

$$P(k/k-1) = F(\hat{\mathbf{x}}(k-1/k-1), k, k-1)P(k-1/k-1)F^{\mathrm{T}}(\ldots) + Q(k-1)$$

Filtering:

$$\hat{\mathbf{x}}(k/k) = \hat{\mathbf{x}}(k/k-1) + \mathbf{k}(k)\{z(k) - g(\hat{\mathbf{x}}(k/k-1), k)\} \tag{5.5}$$

$$P(k/k) = P(k/k-1) - \mathbf{k}(k)\mathbf{h}^{\mathrm{T}}(\hat{\mathbf{x}}(k/k-1), k)P(k/k-1)$$

$$\mathbf{k}(k) = P(k/k-1)\mathbf{h}(\hat{\mathbf{x}}(k/k-1), k)\{\mathbf{h}^{\mathrm{T}}(\ldots)P(k/k-1)\mathbf{h}(\ldots) + R(k)\}^{-1}$$

where (…) means the same quantity between brackets, as used before in the equation. The functions $f()$ and $g()$ are evaluated using the current best estimates for the state. Since the values in the transition matrix $F()$ and measurement vector $\mathbf{h}()$ depend upon the estimates, the matrices $P(k/k-1)$ and $P(k/k)$ cannot be precomputed as for linear prediction and filtering.

In addition, owing to the nature of the assumptions made in the approximated linearization of the model, the extended Kalman filter is no longer optimal and its performance may be much more sensitive to the correctness of the initial guesses $\hat{\mathbf{x}}(0/0)$ and $P(0/0)$. This problem can be alleviated to some degree by iterating more times through the data, each time using as initial guesses the values of the final estimated state in the previous filter pass. However, this works only when the state remains constant in the index variable and in this case offline curve fitting should be considered as an alternative.

5.3   Iterated extended Kalman filter

In certain applications, local iteration may result into performance improvement. The fundamental idea of the iterated filter is that once $\hat{\mathbf{x}}(k/k)$ is generated, then its values would serve as a better estimate for evaluating $g()$ and $\mathbf{h}()$ in the filtering update equations. Then the estimated state after measurement incorporation can be recomputed, iteratively if desired. The iterated extended Kalman filter is given by:

Filtering:

$$\hat{\mathbf{x}}^{i+1}(k/k) = \hat{\mathbf{x}}(k/k-1) + \mathbf{k}^{i}(k)\{z(k) - g(\hat{\mathbf{x}}^{i}(k/k), k)$$

$$-\mathbf{h}^{\mathrm{T}}(\hat{\mathbf{x}}^{i}(k/k), k)[\hat{\mathbf{x}}(k/k-1) - \hat{\mathbf{x}}^{i}(k/k)]\} \tag{5.6}$$

$$\mathbf{k}^i(k) = P(k/k-1)\mathbf{h}(\hat{\mathbf{x}}^i(k/k),k)\{\mathbf{h}^{\mathrm{T}}(\ldots)P(k/k-1)\mathbf{h}(\ldots)+R(k)\}^{-1}$$

where $i = 0,1,\ldots,\ell-1$; initially $\hat{\mathbf{x}}^0(k/k) = \hat{\mathbf{x}}(k/k-1)$ and finally $\hat{\mathbf{x}}(k/k) = \hat{\mathbf{x}}^\ell(k/k)$.

$$\mathbf{h}^{\mathrm{T}}(\hat{\mathbf{x}}^i(k/k),k) = \frac{dg(\mathbf{x}(k),k)}{d\mathbf{x}(k)}\bigg|_{\mathbf{x}(k)=\hat{\mathbf{x}}^i(k/k)} \tag{5.7}$$

The matrix $P(k/k)$ is updated once after the iteration is stopped:

$$P(k/k) = P(k/k-1) - \mathbf{k}^{\ell-1}(k)\mathbf{h}^{\mathrm{T}}(\hat{\mathbf{x}}^{\ell-1}(k/k),k)P(k/k-1) \tag{5.8}$$

Note that $\hat{\mathbf{x}}^1(k/k)$ is just the $\hat{\mathbf{x}}(k/k)$ provided by the extended Kalman filter and when the measurement as well as the system equation are linear the conventional Kalman filter is obtained where iteration offers no improvement.

5.4 Iterated linear filter-smoother

It is also possible to improve for propagation in the index once the measurement $z(k)$ is taken by applying smoothing from k one-step backward in the sequence to $k-1$. Incorporating such a local iteration into the extended Kalman filter structure yields what is termed the iterated linear filter-smoother. We just give the equations:

Prediction:

$$\hat{\mathbf{x}}^i(k/k-1) = f(\hat{\mathbf{x}}^i(k-1/k),k-1) + F(\hat{\mathbf{x}}^i(k-1/k),k,k-1)$$
$$\cdot \{\hat{\mathbf{x}}(k-1/k-1) - \hat{\mathbf{x}}^i(k-1/k)\} \tag{5.9}$$

$$P^i(k/k-1) = F(\hat{\mathbf{x}}^i(k-1/k),k,k-1)P(k-1/k-1)F^{\mathrm{T}}(\ldots) + Q(k-1)$$

Filtering:

$$\hat{\mathbf{x}}^{i+1}(k/k) = \hat{\mathbf{x}}^i(k/k-1) + \mathbf{k}^i(k)\{z(k) - g(\hat{\mathbf{x}}^i(k/k),k)$$
$$-\mathbf{h}^{\mathrm{T}}(\hat{\mathbf{x}}^i(k/k),k)[\hat{\mathbf{x}}^i(k/k-1) - \hat{\mathbf{x}}^i(k/k)]\} \tag{5.10}$$

$$\mathbf{k}^i(k) = P^i(k/k-1)\mathbf{h}(\hat{\mathbf{x}}^i(k/k),k)\{\mathbf{h}^{\mathrm{T}}(\ldots)P^i(k/k-1)\mathbf{h}(\ldots)+R(k)\}^{-1}$$

Smoothing:

$$\hat{\mathbf{x}}^{i+1}(k-1/k) = \hat{\mathbf{x}}(k-1/k-1) + A^i(k-1)\left[\hat{\mathbf{x}}^{i+1}(k/k) - \hat{\mathbf{x}}^i(k/k-1)\right] \tag{5.11}$$

$$A^i(k-1) = P(k-1/k-1)F^{\mathrm{T}}(\hat{\mathbf{x}}^i(k-1/k),k,k-1)P^i(k/k-1)^{-1}$$

where $i = 0,1,\ldots,\ell-1$; initially $\hat{\mathbf{x}}^0(k-1/k) = \hat{\mathbf{x}}(k-1/k-1)$, $\hat{\mathbf{x}}^0(k/k) = \hat{\mathbf{x}}^0(k/k-1)$ and finally $\hat{\mathbf{x}}(k/k) = \hat{\mathbf{x}}^\ell(k/k)$.

$$F(\hat{\mathbf{x}}^i(k-1/k),k,k-1) = \frac{df(\mathbf{x}(k-1),k-1)}{d\mathbf{x}(k-1)}\bigg|_{\mathbf{x}(k-1)=\hat{\mathbf{x}}^i(k-1/k)} \tag{5.12}$$

$$\mathbf{h}^{\mathrm{T}}(\hat{\mathbf{x}}^i(k/k), k) = \frac{dg(\mathbf{x}(k), k)}{d\mathbf{x}(k)}\bigg|_{\mathbf{x}(k)=\hat{\mathbf{x}}^i(k/k)}$$

The matrix $P(k/k)$ is updated once after termination of the iteration:

$$P(k/k) = P^{\ell-1}(k/k-1) - \mathbf{k}^{\ell-1}(k)\mathbf{h}^{\mathrm{T}}(\hat{\mathbf{x}}^{\ell-1}(k/k), k)P^{\ell-1}(k/k-1) \quad (5.13)$$

Note that $\hat{\mathbf{x}}^1(k/k)$ is just the $\hat{\mathbf{x}}(k/k)$ provided by the extended Kalman filter. When the system equation is linear, the prediction-smoothing cycle offers no improvement and the iterated extended Kalman filter is obtained. One can encounter the situation where the system equation is nonlinear with a linear measurement equation. The filtering part of the iterated linear filter-smoother then reduces to:

Filtering:

$$\hat{\mathbf{x}}^{i+1}(k/k) = \hat{\mathbf{x}}^i(k/k-1) + \mathbf{k}^i(k)\{z(k) - \mathbf{h}^{\mathrm{T}}(k)\hat{\mathbf{x}}^i(k/k-1)\} \quad (5.10)$$

$$\mathbf{k}^i(k) = P^i(k/k-1)\mathbf{h}(k)\{\mathbf{h}^{\mathrm{T}}(k)P^i(k/k-1)\mathbf{h}(k) + R(k)\}^{-1}$$

$$\text{and}: \quad P(k/k) = P^{\ell-1}(k/k-1) - \mathbf{k}^{\ell-1}(k)\mathbf{h}^{\mathrm{T}}(k)P^{\ell-1}(k/k-1) \quad (5.13)$$

It should be noted that there is an important difference with the conventional Kalman filter. In all the described extended algorithms for nonlinear estimation, the gain vectors $\mathbf{k}(k)$ are actually random variables, depending on the estimates through the transition matrix $F()$ and measurement vector $\mathbf{h}()$. This results from the fact that we have chosen to linearize $f()$ and $g()$ for the current state. Hence, $\mathbf{k}(k)$ must be computed in real time, it cannot be precomputed. Furthermore, the matrices $P(k/k-1)$, $P^i(k/k-1)$, and $P(k/k)$ become also random, depend on the history and lose statistically their meaning as covariance matrices.

5.5 Nonlinear smoothing

The basic idea of smoothing for nonlinear systems is the same as for the linear case described in Chapter 3, namely, that the filtered estimate $\hat{\mathbf{x}}(k/k)$ can be improved if past measurement data are processed. The short discussion here is confined to fixed-interval smoothing. The smoothed estimate $\hat{\mathbf{x}}(k/N)$ for a linear system can be obtained by pooling of the estimated states provided by a forward and backward running Kalman filter. If the same approach is taken in the nonlinear case, then the pooling equation still holds. Again, by analogy with linear systems, the RTS-algorithm in terms of a forward Kalman filter estimate and a smoothed backward estimate can be applied. The exception is that one has to correct the state equation for a term that is the solution of a nonlinear differential equation. When the system equation is linear, this term equals to zero and one obtains the conventional RTS-smoother. In addition, when for a nonlinear system equation smoothing is performed only one-step backward, the equations pertain to the ones given in (5.11).

5.6 Examples

Example 5.1: Firstly, the same kinetic system as Example 2.4 is considered. Also here, the derivatives in the measurement vector $\mathbf{h}(k)$ for a consecutive first-order reaction are given. The applied transition matrix is the identity matrix $F(k, k-1)=I$ and the system noise covariance matrix the zero matrix $Q(k-1)=0$. For the kinetic system, the reaction rates $k_1=0.1$ and $k_2=0.2$ and the values $x_0=0.05$ and $x_\infty=0.9$ are employed. A measurement noise variance $R(k)=10^{-5}$ is used for simulation and estimation. The extended Kalman filter performs badly. Better results were obtained by the iterated extended Kalman filter operating with $\ell=5$, which are given in Fig. 5.2. The measurements and final fit based on the last estimated state are shown in Fig. 5.1. Both figures demonstrate that the estimated state converges toward constant values and gives a good final estimated fit of the measurements (Figs. 5.1 and 5.2).

Example 5.2: The same chromatographic system as Example 2.5 is considered. Here, the derivatives in the measurement vector $\mathbf{h}(k)$ for chromatography with one peak plus baseline are given. The transition matrix is identity $F(k, k-1)=I$ and the system noise covariance matrix zero $Q(k-1)=0$.

A measurement noise variance $R(k)=10^{-5}$ is used. Further $H=0.8$, $A=0.25$, $\sigma=5$, $t_r=16$, $a_0=0.1$, $a_1=0.004$, and $a_2=-2\cdot10^{-5}$. The measurements used are given in Fig. 5.3. The extended Kalman filter performs badly. The estimation results by the iterated extended Kalman filter with $\ell=5$ are shown in Fig. 5.4. The final estimated fit is also depicted in Fig. 5.3. For plotting $\hat{H}()$, $\hat{A}()$, $\hat{\sigma}()/10$, $\hat{t}_r()/50$, $\hat{a}_0()$, $\hat{a}_1()$, and $\hat{a}_2()$ are used. It demonstrates clearly that the estimated state converges toward constant values and gives a good final

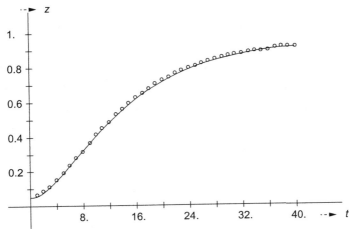

FIG. 5.1   Kinetics: the measurements and final estimated fit by the iterated extended Kalman filter.

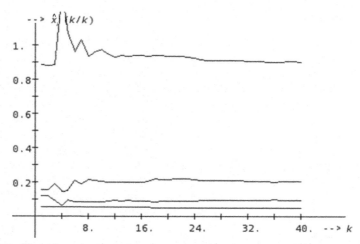

FIG. 5.2   The estimated state by the iterated extended Kalman filter. From above to below at $k = 40$: $\hat{x}_\infty()$, $\hat{k}_2()$, $\hat{k}_1()$, and $\hat{x}_0()$.

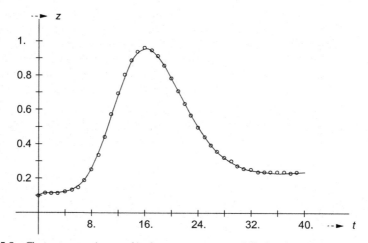

FIG. 5.3   Chromatogram (one peak): the measurements and final estimated fit by the iterated extended Kalman filter.

estimated fit of the measurements. Application of nonlinear estimation for both the three-peak Examples 2.6 and 2.7 of chromatography and spectroscopy was unsuccessful.

Example 5.3: The periodic or oscillatory system of Example 3.2 is considered. The extended state space model with inclusion of the periodic term $\omega(k) \equiv \omega \Delta t$ is described by:

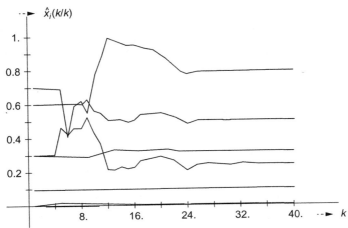

FIG. 5.4 The estimated state by the iterated extended Kalman filter. From above to below at $k=40$: $\hat{H}()$, $\hat{\sigma}()/10$, $\hat{t}_r()/50$, $\hat{A}()$, $\hat{a}_0()$, $\hat{a}_1()$, and $\hat{a}_2()$.

$$\mathbf{x}^T(k) = (x_1(k)\ x_2(k)\ \omega(k))\quad \mathbf{h}^T(k) = (1\ 0\ 0)\qquad (5.14)$$

$$f(\mathbf{x}(k-1), k-1) = \begin{pmatrix} x_1(k-1)\cos\omega(k-1) + x_2(k-1)\sin\omega(k-1) \\ -x_1(k-1)\sin\omega(k-1) + x_2(k-1)\cos\omega(k-1) \\ \omega(k-1) \end{pmatrix}$$

$$F(\mathbf{x}(k-1), k, k-1) = \frac{df(\mathbf{x}(k-1), k-1)}{d\mathbf{x}(k-1)}$$

$$= \begin{pmatrix} \cos\omega(k-1) & \sin\omega(k-1) & x_2(k-1)\cos\omega(k-1) - x_1(k-1)\sin\omega(k-1) \\ -\sin\omega(k-1) & \cos\omega(k-1) & -x_2(k-1)\sin\omega(k-1) - x_1(k-1)\cos\omega(k-1) \\ 0 & 0 & 1 \end{pmatrix}$$

Note that the measurement equation is linear. In Fig. 5.5, the measurements for a simulation with $\omega(k) = \omega\Delta t = 0.3142$ are plotted. The measurement noise variance is $R(k) = 10^{-4}$; the system noise variances are $Q_{11}(k-1) = Q_{22}(k-1) = 10^{-4}$ and $Q_{33}(k-1) = 10^{-8}$. In Fig. 5.6, the estimated state by the iterated linear filter-smoother with $\ell = 3$ is depicted. The extended Kalman filter gives for this example somewhat poorer results.

Example 5.4: The exponential system or first-order process of Example 3.3 is taken. The extended state space model with inclusion of the exponential factor $\Phi(k) = e^{-\Delta t/T_x}$ is described by:

$$\mathbf{x}^T(k) = (x(k)\ x_\infty(k)\ \Phi(k))\quad \mathbf{h}^T(k) = (1\ 0\ 0)\qquad (5.15)$$

$$f(\mathbf{x}(k-1), k-1) = \begin{pmatrix} \Phi(k-1)x(k-1) + (1 - \Phi(k-1))x_\infty(k-1) \\ x_\infty(k-1) \\ \Phi(k-1) \end{pmatrix}$$

$$F(\mathbf{x}(k-1), k, k-1) = \frac{df(\mathbf{x}(k-1), k-1)}{d\mathbf{x}(k-1)}$$

$$= \begin{pmatrix} \Phi(k-1) & 1 - \Phi(k-1) & x(k-1) - x_\infty(k-1) \\ 0 & 1 & 0 \\ 0 & 0 & 1 \end{pmatrix}$$

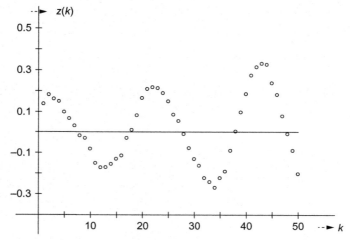

FIG. 5.5   Periodic example: the measurements.

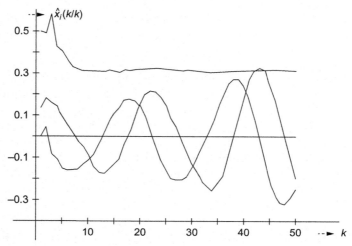

FIG. 5.6   The estimated state by the iterated linear filter-smoother. From above to below at $k = 1$: $\hat{\omega}()$, $\hat{x}_1()$, and $\hat{x}_2()$.

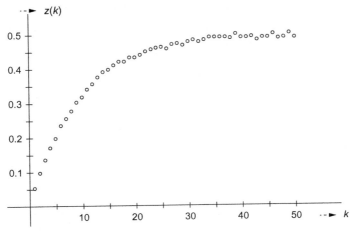

FIG. 5.7 Exponential example: the measurements.

Note that the measurement equation is linear. In Fig. 5.7, $\Phi(k)=0.9$ and $x_\infty(k)=0.5$ are used for simulation of the measurements. The measurement noise variance employed is $R(k)=10^{-5}$; the system noise variances applied are $Q_{11}(k-1)=Q_{22}(k-1)=10^{-6}$ and $Q_{33}(k-1)=10^{-10}$. Fig. 5.8 shows the estimated state by the iterated linear filter-smoother with $\ell=3$. For plotting $\hat{\Phi}(k/k)$ is divided by a factor 2. Here, the extended Kalman filter gives worse results.

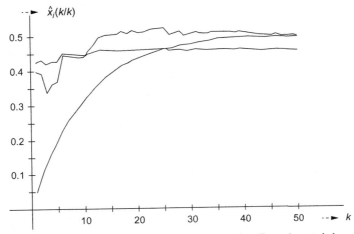

FIG. 5.8 The estimated state by the iterated linear filter-smoother. From above to below at $k=20$: $\hat{x}_\infty()$, $\hat{\Phi}()/2$, and $\hat{x}()$.

Example 5.5: The last complete nonlinear example is taken from literature. Consider the discrete nonlinear process given by the following equations:

$$x(k) = (1 - \theta_1)x(k-1) + \theta_2 x^2(k-1) + w(k) \tag{5.16}$$

$$z(k) = x^2(k) + x^3(k) + v(k)$$

where $x(k)$ is a variable at index k, θ_1, and θ_2 are unknown parameters.

The extended state space model is described by:

$$\mathbf{x}^T(k) = (x(k) \quad \theta_1(k) \quad \theta_2(k)) \tag{5.17}$$

$$g(\mathbf{x}(k), k) = x^2(k) + x^3(k)$$

$$\mathbf{h}^T(\mathbf{x}(k), k) = \frac{dg(\mathbf{x}(k), k)}{d\mathbf{x}(k)} = (2x(k) + 3x^2(k) \quad 0 \quad 0)$$

$$f(\mathbf{x}(k-1), k-1) = \begin{pmatrix} (1 - \theta_1(k-1))x(k-1) + \theta_2(k-1)x^2(k-1) \\ \theta_1(k-1) \\ \theta_2(k-1) \end{pmatrix}$$

$$F(\mathbf{x}(k-1), k, k-1) = \frac{df(\mathbf{x}(k-1), k-1)}{d\mathbf{x}(k-1)}$$

$$= \begin{pmatrix} 1 - \theta_1(k-1) + 2\theta_2(k-1)x(k-1) & -x(k-1) & x^2(k-1) \\ 0 & 1 & 0 \\ 0 & 0 & 1 \end{pmatrix}$$

For the nonlinear system, $\theta_1(k) = 0.0125$ and $\theta_2(k) = 0.01$ are employed. The measurement noise variance used is $R(k) = 10^{-5}$; the system noise variances applied are $Q_{11}(k-1) = 10^{-6}$ and $Q_{22}(k-1) = Q_{33}(k-1) = 10^{-10}$. In Fig. 5.9

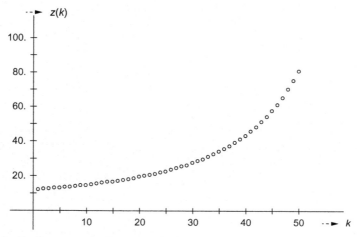

FIG. 5.9 Nonlinear example: the measurements.

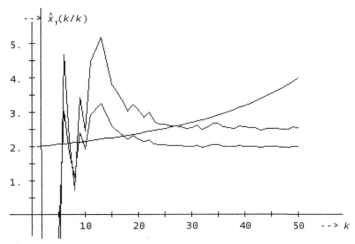

FIG. 5.10   The estimated state by the iterated linear filter-smoother. From above to below at $k = 50$: $\hat{x}()$, $200*\hat{\vartheta}_1()$, and $200*\hat{\vartheta}_2()$.

the measurements are shown, and in Fig. 5.10 the estimated state by the complete iterated linear filter-smoother with $\ell = 3$ is plotted. The values of the constant state variables $\hat{\theta}_1(k/k)$ and $\hat{\theta}_2(k/k)$ are multiplied by a factor 200 for plotting.

Here again, the extended Kalman filter performs badly.

Chapter 6

The multicomponent system

Chapter outline

6.1 Multicomponent analysis

6.1.1 Modeling

Within the recursive context, the model for multicomponent analysis is based on the one from classical least squares simply transformed to the state space model:

$$\mathbf{x}(k) = \mathbf{x}(k-1) \tag{6.1}$$

$$z(k) = \mathbf{h}^{\mathrm{T}}(k)\mathbf{x}(k) + v(k)$$

As example, consider spectroscopy where the law of Lambert-Beer holds, providing a linear and additive measurement equation. Here, the state $\mathbf{x}(k)$ contains as the unknowns the concentrations of the pure components present in the sample. The system equation pinpoints the state as index invariant, i.e., constant in the sequence index, because the transition matrix $F(k, k-1) = I$ and system noise $\mathbf{w}(k-1) = 0$ are vanished. The measurement equation relates the state to the responses. The values of the measurement vector $\mathbf{h}(k)$ are predetermined or known in advance and change when a new measurement $z(k)$ is processed.

The state vector $\mathbf{x}(k)$ containing the unknown concentrations c_i; $i = 1, 2, \ldots,$ $n-1$ is extended with a constant baseline b. The measurement vector $\mathbf{h}(k)$ comprises the pure component responses added with a dummy "1" or:

$$\mathbf{x}^T(k) = (c_1(k) \; c_2(k) \; \dots \; c_{n-1}(k) \; b(k)) \qquad (6.2)$$

$$\mathbf{h}^T(k) = (h_1(k) \; h_2(k) \; \dots \; h_{n-1}(k) \; 1)$$

The observability criterion (Eq. 3.4) is satisfied when all $\mathbf{h}(k)$ are different. For a numerically better condition of the system, a similarity transformation (Eq. 3.3) can be useful.

6.1.2 Prediction

Prediction pertains to Eqs. (3.11) given in Section 3.3 with the transition matrix $F(k,k-1)=I$ and the system noise covariance matrix $Q(k-1)=0$, where I is the identity matrix and 0 the null matrix.

6.1.3 Filtering

Filtering is assigned to Eqs. (3.26) of the Kalman filter in Section 3.4.

6.1.4 Smoothing

Smoothing is not applicable here because the system noise covariance $Q(k)=0$.

6.1.5 Optimization

6.1.5.1 Classical approach

The experimental design affects directly the imprecision of the estimated parameters given by the covariance matrix $P(k)$ in classical least squares. For an optimal design the covariance matrix must be minimized, i.e., the design matrix $H(k)$ with $n \leq k \leq m$ and the noise covariance matrix $\mathbf{R}(k)$ must be chosen such that $P(k)$ is minimized somehow. In first instance, assume that $\mathbf{R}(k)$ is a diagonal matrix with the scalar noise variances $R(i)$, $i=1,2,\dots,k$ all equal on the diagonal. The covariance matrix can then be written as:

$$P(k) = R(k)\{H^T(k)H(k)\}^{-1} \qquad (6.3)$$

The optimal choice of design vectors will be one that makes the covariance matrix smallest in some sense. Some ways of formulating this mathematically:

$$\text{minimize determinant } |\{H^T(k)H(k)\}^{-1}| \quad \text{D-optimality} \qquad (6.4)$$

$$\text{minimize trace } tr\{\{H^T(k)H(k)\}^{-1}\} \quad \text{A-optimality}$$

$$\text{minimize eigenvalue } \lambda_{max}\{\{H^T(k)H(k)\}^{-1}\} \quad \text{E-optimality}$$

$$\text{maximize predictor } \mathbf{h}^T(i)P(i-1)\mathbf{h}(i) \quad \text{G-optimality}$$

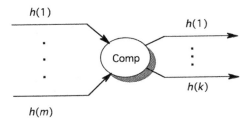

FIG. 6.1 Classical optimization.

An alternative criterion not based on statistical backgrounds is to minimize the condition of a matrix:

$$\kappa(H(k)) = \left\{ \frac{\lambda_{\max}(H^{\mathrm{T}}(k)H(k))}{\lambda_{\min}(H^{\mathrm{T}}(k)H(k))} \right\}^{1/2} \tag{6.5}$$

In Eqs. (6.4) and (6.5), $\lambda_{\min}$ and $\lambda_{\max}$ denote the smallest and largest eigenvalue of the given matrix. The condition $\kappa()$ is a measure of the numerical inaccuracy encountered in the computation of the inverse of a matrix and should only be used for the detection of numerical problems.

The determinant representing the volume of the covariance matrix is used mostly. The computation of the determinant by (6.4) involves the inversion of a matrix. For computational purposes, it is much easier to use the determinant before the inverse is taken. This is justified because of the determinant relation $|A^{-1}| = 1/|A|$ if $|A| \neq 0$. Now, the classical approach to the optimal design problem is to use the determinant as a criterion formulated in the inverse way:

$$\text{Maximize } |P^{-1}(k)| = |H^{\mathrm{T}}(k)\mathbf{R}^{-1}(k)H(k)| \tag{6.6}$$

In statistical literature, the matrix $M(k) = P^{-1}(k)$ is well known as the Fisher information matrix. A representation of the classical optimization based on least squares estimation is shown in Fig. 6.1. Note that the optimal design is computed offline before any measuring is done on the unknown sample.

To choose an optimal target set $H(k)$ from a given dataset $H(m)$ with $n \leq k \leq m$, one has to compute $\frac{m!}{(m-k)!k!}$ determinants for each possible combination of the rows in the target matrix. Of course, this can be an immense computational job. Therefore, the optimal design in a classical sense will only be beneficial for a limited set of design vectors, the target set is small, and when also not too many parameters are involved.

6.1.5.2 Recursive approach

The entropy **H** is defined as the average amount of information given by the probabilities $p(i)$ of all possible mutually exclusive events $\{i\}$ with $\sum_{i=1}^{\ell} p(i) = 1$.

For a discrete set of events, Shannon's formula is applicable by definition:

$$\mathbf{H} = -\sum_{i=1}^{\ell} p(i) ld\{p(i)\} \tag{6.7}$$

According to information theory, the entropy of a multivariate probability density function is defined by replacing the summation by an n-dimensional integral:

$$\mathbf{H} = -\int_{-\infty}^{+\infty} \cdots \int_{-\infty}^{+\infty} p\{\mathbf{x}\} ld(p\{\mathbf{x}\}) d\mathbf{x} \tag{6.8}$$

where $p\{\mathbf{x}\}$ is the probability density function of the n-column random vector $\mathbf{x}$ and $d\mathbf{x}$ is the volume element.

For the Kalman filter, the estimate $\hat{\mathbf{x}}(k/k)$ has a multivariate normal distribution:

$$p\{\mathbf{x}(k)\} = \frac{1}{[(2\pi)^n | P(k/k)|]^{1/2}} e^{-1/2[\mathbf{x}(k) - \hat{\mathbf{x}}(k/k)]^{\mathrm{T}} P^{-1}(k/k)[\mathbf{x}(k) - \hat{\mathbf{x}}(k/k)]} \tag{6.9}$$

with the mean and covariance matrix defined by the integrals:

$$\int_{-\infty}^{+\infty} \cdots \int_{-\infty}^{+\infty} p\{\mathbf{x}(k)\} d\mathbf{x}(k) = 1 \tag{6.10}$$

$$E\{\mathbf{x}(k)\} = \int_{-\infty}^{+\infty} \cdots \int_{-\infty}^{+\infty} \mathbf{x}(k) p\{\mathbf{x}(k)\} d\mathbf{x}(k) = \hat{\mathbf{x}}(k/k)$$

$$E\left\{[\mathbf{x}(k) - \hat{\mathbf{x}}(k/k)][\mathbf{x}(k) - \hat{\mathbf{x}}(k/k)]^{\mathrm{T}}\right\}$$
$$= \int_{-\infty}^{+\infty} \cdots \int_{-\infty}^{+\infty} [\mathbf{x}(k) - \hat{\mathbf{x}}(k/k)][\mathbf{x}(k) - \hat{\mathbf{x}}(k/k)]^{\mathrm{T}} p\{\mathbf{x}(k)\} d\mathbf{x}(k) = P(k/k)$$

Applying formula (6.8) with use of Eqs. (6.9) and (6.10):

$$\mathbf{H}(k) = -\int_{-\infty}^{+\infty} \cdots \int_{-\infty}^{+\infty} p\{\mathbf{x}(k)\}[-1/2 ld\{(2\pi)^n | P(k/k)|\}$$

$$-1/2 ld\{e\}[\mathbf{x}(k) - \hat{\mathbf{x}}(k/k)]^{\mathrm{T}} P^{-1}(k/k)[\mathbf{x}(k) - \hat{\mathbf{x}}(k/k)]]d\mathbf{x}(k) \tag{6.11}$$

$$= 1/2 ld\{(2\pi)^n | P(k/k)|\} + 1/2 ld\{e\} tr\{P^{-1}(k/k) P(k/k)\}$$

where we used the trace relation $\mathbf{x}^{\mathrm{T}} A \mathbf{x} = tr\{A \mathbf{x} \mathbf{x}^{\mathrm{T}}\}$.

Thus, the entropy of a multivariate normal distribution is given by:

$$\mathbf{H}(k) = 1/2ld\{(2\pi e)^n |P(k/k)|\} \tag{6.12}$$

The information yield $I(k)$ is defined as the change in entropy by processing a new measurement:

$$I(k) = \mathbf{H}_{before} - \mathbf{H}_{after} = \mathbf{H}(k-1) - \mathbf{H}(k) \tag{6.13}$$

$$= 1/2ld\{|P(k-1/k-1)|/|P(k/k)|\}$$

$$= 1/2ld\{|P(k-1/k-1)|/|P(k/k-1)|\}$$

$$+ 1/2ld\{|P(k/k-1)|/|P(k/k)|\}$$

$$= I_{pre}(k) + I_{fil}(k)$$

$I_{pre}(k)$ describes the information loss by the prediction part of the Kalman filter. This term is in the case of multicomponent analysis equal to zero since the state is index invariant and there is no system noise. $I_{fil}(k)$ is the information gain for processing the kth measurement by the filtering part of the Kalman filter.

With use of the determinant product rule for the ratio $|P(k/k-1)|/|P(k/k)|$, one gets in first instance the information yield for filtering:

$$I_{fil}(k) = -1/2ld\{|I - \mathbf{k}(k)\mathbf{h}^T(k)|\} \tag{6.14}$$

However, the full use of the recursive properties is confirmed by another important mathematical result (2.35) from Section 2.4, the matrix determinant lemma:

$$|P(k/k-1)| = \{\mathbf{h}^T(k)P(k/k-1)\mathbf{h}(k)/R(k) + 1\}|P(k/k)| \tag{6.15}$$

Incorporating this result into (6.13) gives the information yield in the form:

$$I_{fil}(k) = 1/2ld\{\mathbf{h}^T(k)P(k/k-1)\mathbf{h}(k)/R(k) + 1\} \tag{6.16}$$

This formulation for the information yield takes into account which data points have been previously expressed through the inclusion of the predicted covariance matrix $P(k/k-1)$. Because the term between curly brackets is a scalar, the determinant computation is no longer required, which minimizes the computational effort. The term in Eq. (6.16) consists of two main parts: the covariance matrix $P(k/k-1)$ involved with the "*old knowledge*" and the measurement vector $\mathbf{h}(k)$ and noise variance $R(k)$, supplying the "*new information.*" In addition, noisy measurements have low information yields (high $R(k)$). Theoretically, it expresses the relation between the covariance matrix obtained by the previous experimental design, evaluating the gain of the current measurement vector and the noise variance.

For k to infinity, the Kalman gain vector $\mathbf{k}(k)$ in Eq. (6.14) converges to the null vector, thus the limiting information yield is here:

$$\lim_{k\to\infty} I(k) = \lim_{k\to\infty} I_{fil}(k) = -1/2ld\{|I|\} = 0 \tag{6.17}$$

This agrees with the expectation that for an increasing number of measurements the total information yield $I_{tot}(k)$ reaches toward infinity, i.e., the estimation algorithm knows the estimate $\hat{\mathbf{x}}(k/k)$ as $\mathbf{x}(k)$ exactly and finally no more information $I(k)$ can be obtained. Reformulating the entropy $\mathbf{H}(k)$, this gives for optimization:

$$\text{Minimize} \quad \mathbf{H}(k) = 1/2ld\{(2\pi e)^n | P(k/k)| \} \tag{6.18}$$

$$\text{Maximize} \quad I_{tot}(k) = 1/2ld\{| P(0/0)| / | P(k/k)| \} = I_{tot}(k-1) + I(k)$$

$$\text{Maximize} \quad I(k) = 1/2ld\{\mathbf{h}^{\mathrm{T}}(k)P(k/k-1)\mathbf{h}(k)/R(k) + 1\}$$

$$\text{Maximize} \quad \mathbf{h}^{\mathrm{T}}(k)P(k/k-1)\mathbf{h}(k)/R(k)$$

The optimal design problem is redefined as a sequence of maximized experimental choices with the information yield as criterion. The $m*n$ design matrix $H(m)$ contains the molar spectra plus baseline. A chosen measurement vector is blocked in the design matrix, so it cannot be used again. For the choice of a new $\mathbf{h}(k)$, the design dataset $\{\mathbf{h}(i), R(i)\}$, $i = 1,2,\ldots,\ell$ with $\ell \leq m$ has to be processed only once. This means for each k the computation of ℓ times the final equation (6.18); calculation of the determinant is no longer involved. Because every additionally maximization of the information yield is based on a previously minimized covariance matrix, this procedure is more concisely known as minimax optimization. The minimax criterion is the quintessence of the recursive optimal design computation.

In Fig. 6.2, a representation of the optimization procedure is depicted in case all the measurement noise variances are equal with $\ell \leq m$. The assumption of equal valued noise variances is not necessary. If the noise variance depends on the measurement, the computation can only be done in an online fashion. Based on the current estimate and the measurement vector of interest, it is possible to predict the measurement with its expected noise variance. After a measurement vector is chosen, the measurement with the correct noise variance is used for updating the estimated state and covariance matrix.

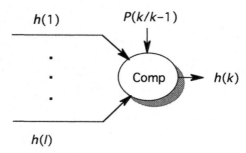

FIG. 6.2 Recursive optimization.

If equal noise variances are assumed in the measurement process, some general properties can be formulated. If the transition matrix $F(k,k-1)$ is identity and the system noise covariance matrix $Q(k-1)$ null, from recursive least squares follows:

$$P(k/k) = \left\{ P^{-1}(0/0) + \sum_{i=1}^{k} \mathbf{h}(i)R^{-1}(i)\mathbf{h}^{\mathrm{T}}(i) \right\}^{-1} \qquad (6.19)$$

One may compute offline an optimal design before any measuring is done on the unknown sample. Each time a measurement vector is chosen, the matrix $P(k/k)$ is updated and the next maximization is repeated. The chosen measurement vectors are blocked in the design matrix. Alternatively, the algorithm may have the freedom to reuse previously chosen measurement vectors again. Let us assume that this is the case. The first maximizations are of course suboptimal, because of the initial matrix $P(0/0)$. The effect of $P(0/0)$ decreases incrementing upon k, whereas the influence of the summation term increases. In the iteration, the algorithm tunes itself up to the minimal determined optimal solution and it can be described as a self-optimizing process. After a number of iterations, the maximized solution repeats itself and equals the classical determined optimal design. Extension to any overdetermined optimal design could be done easily after computation of the determined optimal design. The chosen measurement vectors are now blocked in the design matrix after each next maximization. The computed $P(k/k)$ is no longer a covariance matrix, which can only be computed after starting blocking of the measurement vectors in the design matrix.

An interesting property one can deduce from (6.19) is permutation freedom. The summation term is invariant under permutation of the measurement vectors $\mathbf{h}(i)$ and the corresponding noise variances $R(i)$. When the same measurement vectors are used, this means that the experiment could be performed in any sequential order. This conclusion is important when time-dependent effects are of interest.

The optimal design computation is invariant under a similarity transformation:

$$\mathbf{h}^{*\mathrm{T}}(k) = \mathbf{h}^{\mathrm{T}}(k)T(k) \qquad \mathbf{x}^{*}(k) = T^{-1}(k)\mathbf{x}(k) \qquad (6.20)$$

$$P^{*}(k/k) = T^{-1}(k)P(k/k)T^{-\mathrm{T}}(k)$$

The determinant of the covariance matrix $|P(k/k)|$ is divided by the constant $|T(k)|^{2}$ into the new one $|P^{*}(k/k)|$. The optimal design computation remains unaffected.

A relative measurement vector is the simplest case of a similarity transformation.

Another result is that the optimal design computation is invariant for an additive factor or baseline b_i on every coefficient $h_i(k)$, $i = 1,2,\ldots,n-1$ in the design matrix, provided an extra component $\mathbf{h}^{\mathrm{T}}(k) = (h_1(k) \quad h_2(k) \quad \ldots \quad h_{n-1}(k)\ 1)$

is included. Compared to a measurement vector with no dummy "1" the computed optimal design will be different. Also, the estimated state will be different.

6.1.6 Adaptation

Until now, it is always assumed that the measurement noise variance $R(k)$ is known a priori. In practice, however, this is not always the case. The adaptation problem pinpoints to the search for the right algorithm to assess the measurement noise variance, which directly influences the progress of the integrated estimation procedures implemented. It will be assumed that the noise variance is independent of the measurement and stays constant with a changing index k. One may presume that for the next sample to be analyzed the noise variance does not change. Serious theoretical difficulties are encountered when the noise variance is a function of the measurement $z(k)$ and/or the measurement vector $\mathbf{h}(k)$ is subject to noise.

In first instance, one can use a preset measurement noise variance $R(k)$ in the Kalman filter and compute it offline after all the measurements have been processed. One just uses the residuals based on the last filtered state in the unbiased equation:

$$\hat{R}(N) = \frac{1}{N-n} \sum_{k=1}^{N} \left\{ z(k) - \mathbf{h}^{\mathrm{T}}(k)\hat{\mathbf{x}}(N/N) \right\}^2 \tag{6.21}$$

and the covariance matrix $P^*(N/N)$ is corrected afterward according to:

$$P^*(N/N) = P(N/N)\hat{R}(N)/R(N) \tag{6.22}$$

It is still possible to implement the adaptation of the noise variance online by integrating it in the recursive estimation procedure. Because the estimate $\hat{\mathbf{x}}(k/k)$ depends on the noise variance $R(k)$, it is not possible to use the adapted $\hat{R}(k)$ directly. One has to use its prediction $\hat{R}(k-1)$ in the Kalman filter resulting in suboptimal online estimation. Thus, the Kalman gain vector becomes:

$$\mathbf{k}(k) = P(k/k-1)\mathbf{h}(k)\left\{ \mathbf{h}^{\mathrm{T}}(k)P(k/k-1)\mathbf{h}(k) + \hat{R}(k-1) \right\}^{-1} \tag{6.23}$$

Online adaptation is either based on the innovations or the residuals that are corrected for their theoretical expected values by means of covariance matching.

The commonly employed formula for adaptation based on the innovations does not work properly in practice, so this equation shall not be given. Firstly, a possible equation for adaptation of the noise variance follows from the residuals:

$$\hat{R}(k) = \frac{1}{k} \sum_{i=1}^{k} \left(z(i) - \mathbf{h}^{\mathrm{T}}(i)\hat{\mathbf{x}}(i/i) \right)^2 + \mathbf{h}^{\mathrm{T}}(i)P(i/i)\mathbf{h}(i) \tag{6.24}$$

Rewriting gives after some manipulations simply the following recursive estimator to be used for online adaptation:

$$\hat{R}(k) = (1 - 1/k)\hat{R}(k-1) + 1/k\left\{\left(z(k) - \mathbf{h}^T(k)\hat{\mathbf{x}}(k/k)\right)^2 + \mathbf{h}^T(k)P(k/k)\mathbf{h}(k)\right\}$$

(6.25)

The residual and correction term needs some extra online computation beyond the ones for the Kalman filter. For practical reasons, it is assumed that the adapted measurement noise variance $\hat{R}(k)$ is uniformly distributed between a lower limit $\underline{R}$ and an upper limit $\overline{R}$, which results in boundary conditions to be chosen:

$$\hat{R}(k) < \underline{R} \Rightarrow \hat{R}(k) = \underline{R}$$

(6.26)

$$\hat{R}(k) > \overline{R} \Rightarrow \hat{R}(k) = \overline{R}$$

In this way, one can anticipate on a possible divergence of the Kalman filter.

The second and more complex recursive estimator for online adaptation is derived from the concept of fixed-point smoothing operating on the innovations $v(k)$, which is a special case of the results presented later.

We just give Eqs. (6.42) for $Q(k-1)=0$, $F(k,k-1)=I$ and $B(k,k-1)=I$:

$$\hat{R}(k) = (1 - 1/k)\hat{R}(k-1) + 1/k\{v^2(k) + \mathbf{h}^T(k)P(k/k-1)\mathbf{h}(k)$$

$$-2[\mathbf{k}(k)v(k)]^T C_R(k) + tr\{T(k)S_R(k)\}\}$$

(6.27)

$$T(k) = \mathbf{k}(k)v(k)v^T(k)\mathbf{k}^T(k) + P(k/k) - P(k/k-1)$$

$$C_R(k) = \mathbf{h}(k)v(k) + C_R(k-1) - S_R(k-1)\mathbf{k}(k-1)v(k-1)$$

$$C_R(0) = 0$$

$$S_R(k) = \mathbf{h}(k)\mathbf{h}^T(k) + S_R(k-1)$$

$$S_R(0) = 0$$

$$\hat{R}(k) < \underline{R} \Rightarrow \hat{R}(k) = \underline{R}$$

$$\hat{R}(k) > \overline{R} \Rightarrow \hat{R}(k) = \overline{R}$$

Also here, it is assumed that the adapted measurement noise variance $\hat{R}(k)$ is uniformly distributed between the boundaries $\underline{R}$ and $\overline{R}$, which is a reasonable assumption in practice.

If necessary, a fading memory factor can be incorporated in the online adaptation algorithms. In this way, one can adjust to an unknown changing noise variance. In analytical practice, we always have to deal with a limited number of measurements so this possibility was not implemented.

It should be noted that for online adaptation the estimated state $\hat{\mathbf{x}}(k/k)$ is statistically still valid, but the matrix $P(k/k)$ loses its meaning as covariance matrix and cannot be used for evaluation and/or control purposes.

6.1.7 Evaluation

The concentrations of an unknown sample are directly involved in the estimated state $\hat{\mathbf{x}}(k/k)$ with its covariance matrix $P(k/k)$. Student's t confidence limits (Eq. 4.10) as given in Chapter 4 on statistics hold in case of a correct preset measurement noise variance $R(k)$ online or the adapted $\hat{R}(N)$ offline (Eq. 6.21) after all the measurements have been processed. For a correct preset measurement noise variance, use the z-value $z(1-\alpha/2)=t(1-\alpha/2,\infty)$.

The χ^2-confidence limits based on $\hat{R}(N)$ for the measurement noise variance (Eq. 2.17) are also valid in case of offline adaptation. Until now, no formula is known for the confidence limits of the measurement noise variance adapted online by either Eqs. (6.25) or (6.27). When the measurement noise variance is adapted online, the estimated state $\hat{\mathbf{x}}(k/k)$ is statistically still valid, but the matrix $P(k/k)$ loses its meaning as covariance matrix and thus cannot be used for evaluation purposes. The only way to get the right covariance matrix is to run the Kalman filter once again with a preset measurement noise variance adapted from the last previous run. Thereafter, the confidence limits can be employed again.

6.1.8 Control

Here, one can use as a stopping criterion for the covariance $P(k/k)$ diagonal elements:

$$\text{if max} \quad t(1-\alpha/2, k-n)\sqrt{P_{ii}(k/k)} \leq S_{crit} \quad i=1,2,\dots,n-1 \tag{6.28}$$

$$\text{then stop else continue}$$

where S_{crit} is the critical imprecision to be chosen as a proper analytical goal.

This equation only holds in case of a correct preset noise variance $R(k)$ in the online estimation algorithm and Student's t value reduces to $z(1-\alpha/2)=t(1-\alpha/2,\infty)$.

It should be noted that now the measurement vectors as well as the number of measurements to perform can be precomputed offline by recycling through the filtering, optimization, and control algorithms before any measuring on the sample is done. If the measurement noise variance is adapted online, the matrix $P(k/k)$ is no covariance matrix anymore and cannot be used in the control algorithm.

6.1.9 Restriction

Suppose the unknown state $\mathbf{x}(k)$ is estimated by $\hat{\mathbf{x}}(k/k)$ given the observations $Z(k)=\{z(1),z(2),\dots,z(k)\}$ and the state is subject to a linear inequality constraint:

$$L\mathbf{x}(k) \leq \mathbf{c} \tag{6.29}$$

where L is not necessarily a square $p*n$ matrix, such that the $p*p$ inverse matrix $(LL^T)^{-1}$ exists and $\mathbf{c}$ is some p-column constraint vector. The problem is to find an estimate $\hat{\mathbf{x}}^*(k/k)$ based on $\hat{\mathbf{x}}(k/k)$, which will minimize the mean square error, subject to the inequality constraint. The cost criterion $J(k)$ to minimize is defined as:

$$J(k) = [\hat{\mathbf{x}}^*(k/k) - \hat{\mathbf{x}}(k/k)]^T[\hat{\mathbf{x}}^*(k/k) - \hat{\mathbf{x}}(k/k)] + 2\ell^T(k)\{\mathbf{c} - L\hat{\mathbf{x}}^*(k/k)\} \quad (6.30)$$

where $\ell(k)$ is the p-column Lagrange multiplier.

Differentiation of the cost criterion $J(k)$ with respect to $\hat{\mathbf{x}}^*(k/k)$, setting to zero and taking the transpose readily provide:

$$2[\hat{\mathbf{x}}^*(k/k) - \hat{\mathbf{x}}(k/k)] - 2L^T\ell(k) = 0 \quad (6.31)$$

$$\text{or}: \quad \hat{\mathbf{x}}^*(k/k) = \hat{\mathbf{x}}(k/k) + L^T\ell(k) \quad (6.32)$$

$$\ell(k) = (LL^T)^{-1}\{\mathbf{c} - L\hat{\mathbf{x}}(k/k)\}$$

Rewriting gives finally the correction equations for the inequality constraint:

$$\hat{\mathbf{x}}^*(k/k) = \hat{\mathbf{x}}(k/k) + \lambda(k) \quad (6.33)$$

$$P^*(k/k) = P(k/k) + \lambda(k)\lambda^T(k)$$

$$\lambda(k) = L^T(LL^T)^{-1}\{\mathbf{c} - L\hat{\mathbf{x}}(k/k\}$$

where $\lambda(k)$ is the n-column correction vector. Note that the same result can be derived for estimation in general, i.e., prediction, filtering as well as smoothing.

The inequality constraint in multicomponent analysis reads as: the concentrations are at least zero. In this case $L = (I_{n-1} \, 0_{n-1})$ and $\mathbf{c} = 0_{n-1}$, where I_{n-1} is the $(n-1)*(n-1)$ identity matrix and 0_{n-1} the $(n-1)$-column null vector. It follows that every element in the estimate $\hat{\mathbf{x}}(k/k)$, which is less than zero, should be set equal to zero in the presence of the inequality constraint. Thereafter, the correction vector $\lambda(k)$ (6.33) is adjusted and used later to correct the covariance matrix. Every element in the estimate $\hat{\mathbf{x}}(k/k)$ larger than or equal to zero is left unchanged. In case of stochastic drift, use $L = (I_{n-2} \, 0_{n-2} \, 0_{n-2})$ and $\mathbf{c} = 0_{n-2}$.

For restriction, the information $I_{res}(k)$ added to the information yield $I(k)$ gives:

$$I(k) = I_{pre}(k) + I_{fil}(k) + I_{res}(k) \quad (6.34)$$

$$I_{res}(k) = -1/2ld\{|I + \lambda(k)\lambda^T(k)P^{-1}(k/k)|\} \quad (6.35)$$

Note that $I_{res}(k)$ purveys an information loss because the determinant term between curly brackets of the $ld\{\}$ is always equal or greater than 1.

6.2 Stochastic drift

6.2.1 Modeling

For multicomponent analysis, where the measurements are corrupted by measurement noise as well as stochastic drift, the general state space model is valid:

$$\mathbf{x}(k) = F(k, k-1)\mathbf{x}(k-1) + \mathbf{w}(k-1) \tag{6.36}$$

$$z(k) = \mathbf{h}^{\mathrm{T}}(k)\mathbf{x}(k) + v(k)$$

For the stochastic drift model the transition matrix, state vector, measurement vector, and system noise vector are described by:

$$F(k, k-1) = \begin{pmatrix} I_{n-2} & 0_{n-2,2} \\ 0_{n-2,2}^{\mathrm{T}} & 1 & 1 \\ & 0 & 1 \end{pmatrix} \tag{6.37}$$

$$\mathbf{x}^{\mathrm{T}}(k) = (c_1(k) \ \ c_2(k) \ \ \ldots \ \ c_{n-2}(k) \ \ d(k) \ \ \alpha(k))$$
$$\mathbf{h}^{\mathrm{T}}(k) = (h_1(k) \ \ h_2(k) \ \ \ldots \ \ h_{n-2}(k) \ \ 1 \ \ 0)$$
$$\mathbf{w}^{\mathrm{T}}(k-1) = (0_{n-2}^{\mathrm{T}} \ \ w_{n-1}(k-1) \ \ w_n(k-1))$$

where $0_{n-2,2}$ is the $(n-2)*2$ zero matrix, I_{n-2} is the $(n-2)*(n-2)$ identity matrix, and 0_{n-2} is the $(n-2)$-column null vector. Note that the new model is just the model of the pure components extended with the two-dimensional stochastic drift model, $d(k)$ and $\alpha(k)$ are the added drift parameters.

The incorporation of stochastic drift in the state space model offers new applications for the compensation of unknown impurities or when the pure component spectra deviate somewhat in the measured sample. This cannot be done by classical least squares. In first instance, it is assumed that the unknown disturbance is smooth in the sequence dependent variable. This can be relaxed somewhat through manipulation of the diagonal elements in the system noise covariance matrix.

6.2.2 Prediction

Prediction is done by Eqs. (3.11) of the Kalman filter in Section 3.3.

6.2.3 Filtering

Filtering is assigned to Eqs. (3.26) of the Kalman filter in Section 3.4.

6.2.4 Smoothing

Since the system noise covariance matrix is not the null matrix, smoothing can be done to improve the estimated stochastic drift offline afterward. Eqs. (3.28) in Section 3.6.1 for fixed-interval smoothing are of particular interest.

6.2.5   Optimization

Cannot be done because there is a stringent relation in the state between $k-1$ and k through the dynamics of the system, described by the transition matrix $F(k,k-1)$ and system noise $\mathbf{w}(k-1)$. In general, the system and measurement noise variances $Q(k-1)$ and $R(k)$ should be minimized for an optimal performance.

6.2.6   Adaptation

It follows that for the operation of the Kalman filter the measurement noise variance $R(k)$ and the system noise covariance matrix $Q(k-1)$ have to be known a priori for all k. We are primarily interested in online adaptation methods. This means that if we drop the assumption that the noise variances are known in advance, we shall have to find an algorithm that simultaneously estimates the state and the noise statistics.

The following additional assumption is made:

$$R(k) = R \quad k = 1, 2, \ldots \tag{6.38}$$

$$Q(k-1) = Q$$

Hence, it is assumed that the noise processes are constant and stationary.

The derivation of online adaptation is rather complicated and only an outline shall be given. A suboptimal recursive estimate of the state is obtained by substituting the predictions $\hat{R}(k-1)$ and $\hat{Q}(k-1)$ for the true values $R(k)$ and $Q(k-1)$ in the Kalman filter:

$$P(k/k-1) = F(k, k-1)P(k-1/k-1)F^{\mathrm{T}}(k, k-1) + \hat{Q}(k-1) \tag{6.39}$$

$$\mathbf{k}(k) = P(k/k-1)\mathbf{h}(k)\left\{\mathbf{h}^{\mathrm{T}}(k)P(k/k-1)\mathbf{h}(k) + \hat{R}(k-1)\right\}^{-1}$$

Online adaptation is divided into two main parts: one for the measurement noise variance and the other for the system noise variances. In principle, the concept of fixed-point smoothing is used. Given all the available measurements $Z(k)$, the estimate of the scalar measurement noise variance can be evaluated as:

$$\hat{R}(k) = 1/k \sum_{i=1}^{k} \hat{v}(i/k)\hat{v}^{\mathrm{T}}(i \mid k) \tag{6.40}$$

with : $\hat{v}(i \mid k) = E\{v(i) \mid Z(k)\}$

One obtains the recursive relation for $\hat{v}(i \mid k)$ with respect to k as:

$$\hat{v}(i \mid k) = \hat{v}(i \mid k-1) - \mathbf{h}^{\mathrm{T}}(i)B(k, i)\mathbf{k}(k)v(k) \tag{6.41}$$

$$B(k, i) = \prod_{j=i}^{k-1} A(j) = B(k-1, i)A(k-1) = B(k-1, i)B(k, k-1)$$

$$A(j) = P(j/j)F^{\mathrm{T}}(j+1,j)P^{-1}(j+1/j)$$

$\hat{v}(k|k-1)$ is equal by definition to the innovation $v(k)$ and $B(i,i)=I$ for all $i \geq 0$. Filling in Eq. (6.41) in (6.40) gives after some manipulations and performing covariance matching afterward a recursive estimator for the measurement noise variance $\hat{R}(k)$, which can be used for online adaptation:

$$\hat{R}(k) = (1 - 1/k)\hat{R}(k-1) + 1/k\{v^2(k) + \mathbf{h}^{\mathrm{T}}(k)P(k/k-1)\mathbf{h}(k)$$

$$-2[\mathbf{k}(k)v(k)]^{\mathrm{T}}C_R(k) + tr\{T(k)S_R(k)\}\} \tag{6.42}$$

$$T(k) = \mathbf{k}(k)v(k)v^{\mathrm{T}}(k)\mathbf{k}^{\mathrm{T}}(k) + P(k/k) - P(k/k-1)$$

$$B(k,k-1) = P(k-1/k-1)F^{\mathrm{T}}(k,k-1)P^{-1}(k/k-1)$$

$$C_R(k) = \mathbf{h}(k)v(k) + B^{\mathrm{T}}(k,k-1)\{C_R(k-1) - S_R(k-1)\mathbf{k}(k-1)v(k-1)\}$$

$$C_R(0) = 0$$

$$S_R(k) = \mathbf{h}(k)\mathbf{h}^{\mathrm{T}}(k) + B^{\mathrm{T}}(k,k-1)S_R(k-1)B(k,k-1)$$

$$S_R(0) = 0$$

$$\hat{R}(k) < \underline{R} \Rightarrow \hat{R}(k) = \underline{R}$$

$$\hat{R}(k) > \overline{R} \Rightarrow \hat{R}(k) = \overline{R}$$

Here, $C_R(k)$ is an n-column vector; $S_R(k)$, $T(k)$, and $B(k,k-1)$ are $n*n$ matrices.

It is assumed that the adapted noise variance $\hat{R}(k)$ is uniformly distributed between the boundaries $\underline{R}$ and $\overline{R}$, which is a reasonable assumption in practice.

Similarly, from the measurements $Z(k)$, an estimate for the system noise covariance matrix will be evaluated as:

$$\hat{Q}(k) = 1/k\sum_{i=1}^{k}\hat{\mathbf{w}}(i-1|k)\hat{\mathbf{w}}^{\mathrm{T}}(i-1|k) \tag{6.43}$$

with : $\hat{\mathbf{w}}(i-1|k) = E\{\mathbf{w}(i-1)|Z(k)\}$

One obtains the recursive relation for $\hat{\mathbf{w}}(i-1/k)$ with respect to k as:

$$\hat{\mathbf{w}}(i-1/k) = \hat{\mathbf{w}}(i-1/k-1) + M(i-1)B(k,i)\mathbf{k}(k)v(k) \tag{6.44}$$

$$M(i-1) = Q(i-1)P^{-1}(i/i-1)$$

$\mathbf{w}(k-1)$ is independent of the measurements $Z(k-1)$, thus $\hat{\mathbf{w}}(k-1/k-1) = 0$ for all $k \geq 1$. Using Eq. (6.44) in (6.43), the estimator of the system noise covariance matrix $\hat{Q}(k)$ is transformed similarly as for $\hat{R}(k)$. In practice, we are only interested in diagonal elements $\hat{Q}_{rr}(k)$ of the system noise covariance matrix $\hat{Q}(k)$. This results into the following recursive estimator for online adaptation:

$$\hat{Q}_{rr}(k) = \hat{Q}_{rr}(k-1) + 1/k \left\{ 2[B(k, k-1)\mathbf{k}(k)v(k)]^{\mathrm{T}} C_{Q_{rr}}(k-1) + tr\{T(k)S_{Q_{rr}}(k)\} \right\}$$

$$M(k-1) = \hat{Q}(k-1)P^{-1}(k/k-1) \tag{6.45}$$

$$C_{Q_{rr}}(k-1) = B^{\mathrm{T}}(k-1, k-2)C_{Q_{rr}}(k-2) + S_{Q_{rr}}(k-1)\mathbf{k}(k-1)v(k-1)$$

$$C_{Q_{rr}}(0) = 0$$

$$S_{Q_{rr}}(k) = M^{\mathrm{T}}(k-1)\mathbf{e}_r \mathbf{e}_r^{\mathrm{T}} M(k-1) + B^{\mathrm{T}}(k, k-1)S_{Q_{rr}}(k-1)B(k, k-1)$$

$$S_{Q_{rr}}(0) = 0$$

$$\hat{Q}_{rr}(k) < \underline{Q}_{rr} \Rightarrow \hat{Q}_{rr}(k) = \underline{Q}_{rr}$$

$$\hat{Q}_{rr}(k) > \overline{Q}_{rr} \Rightarrow \hat{Q}_{rr}(k) = \overline{Q}_{rr}$$

$\mathbf{e}_r$ is an n-column unit vector, $C_{Q_{rr}}(k-1)$ is an n-column vector, $S_{Q_{rr}}(k)$, and $M(k-1)$ are $n*n$ matrices. $T(k)$ and $B(k,k-1)$ are defined in Eq. (6.42). It is assumed that the adapted system noise variances $\hat{Q}_{rr}(k)$ are uniformly distributed within the outer limits $\underline{Q}_{rr}$ and $\overline{Q}_{rr}$, which is a reasonable assumption in practice. For every desired diagonal element $\hat{Q}_{rr}(k)$, Eqs. (6.45) have to be processed again. The algorithm can be modified to adapt off-diagonal elements of the system noise covariance matrix $\hat{Q}(k)$.

If necessary, a fading memory factor can be included in the online adaptation algorithms. In this way, one can adjust to an unknown changing noise environment.

6.2.7 Evaluation

The concentrations of an unknown sample are directly involved in the estimated state $\hat{\mathbf{x}}(k/k)$ with its covariance matrix $P(k/k)$. Student's t confidence limits with $z(1-\alpha/2) = t(1-\alpha/2, \infty)$ in Eq. (4.10) hold in case of a correct preset system noise covariance matrix $Q(k-1)$ and measurement noise variance $R(k)$.

It should be noted that the inclusion of stochastic drift with a nonzero system noise covariance matrix $Q(k-1)$; in particular the diagonal elements $Q_{n-1, n-1}(k-1)$ and $Q_{n, n}(k-1)$; increases all the diagonal elements of the covariance matrix $P(k/k)$. This means that the confidence limits for the estimated concentrations become enlarged in comparison with a system with no stochastic drift. Hence, there is upward a statistical limit toward the application of stochastic drift. Until now, no formula is known for the confidence limits of the adapted system and measurement noise variances. When the system and measurement noise variances are adapted online by the given equations (6.42) and (6.45), the estimated state $\hat{\mathbf{x}}(k/k)$ is statistically still valid, but the matrix $P(k/k)$ loses its meaning as covariance matrix and thus cannot be used for evaluation purposes. The only way to get the right covariance matrix to determine the confidence limits is to run the Kalman filter once again with preset system and measurement noise variances adapted from the last previous run. In case of fixed-interval smoothing, $\hat{x}(k/N)$ and $P(k/N)$ are used.

6.2.8 Control

This is almost similar as described previously in Section 6.1.8.

6.2.9 Restriction

This is the same as described before in Section 6.1.9.

6.3 Examples

Example 6.1: The same multicomponent system as in Example 2.1 is considered. For the state space model using four components plus baseline, see further Section 6.1.1. Serious problems were encountered when restriction is applied in combination with adaptation. Because we value adaptation more than the application of restriction, the last one has been omitted in this example, leaving it as a theoretical artifact. The molar spectra of the pure components employed are depicted in Fig. 6.3. The measurements corrupted by measurement noise with a variance $R(k) = 10^{-5}$ are shown in Fig. 6.4. The estimated state by the Kalman filter for forward scanning, backward scanning, and the optimal selection are plotted in Figs. 6.5–6.7, respectively. It demonstrates clearly that the Kalman filter can track quite well the unknown concentrations toward their correct values. The concentrations should be 0.2, 0.4, 0.6, 0.8, and the baseline 0.1.

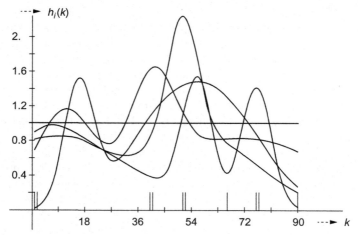

FIG. 6.3 Multicomponent system: the molar spectra involved plus baseline. *Small lines* below denote optimal design selection.

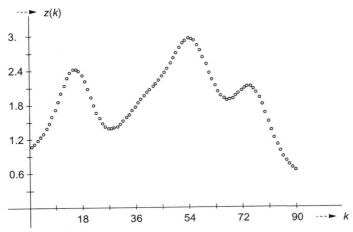

FIG. 6.4 The measurements corrupted by noise.

The total information yields for the applied designs are shown in Fig. 6.8. In case of the optimal selection, the convergence is fastest as should be. The small lines below in Fig. 6.3 denote the optimal design for the first 10 wavelengths chosen for measurement.

Let us now focus on the adaptation of the measurement noise variance $\hat{R}(k)$ starting from $\hat{R}(0) = 5 \cdot 10^{-5}$. For the adapted measurement noise variance similar graphs as before can be produced for state estimation, which shall not be given. In Fig. 6.9, the adaptation progress by Eq. (6.25) is displayed for the three

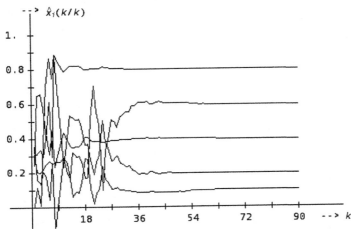

FIG. 6.5 The estimated state by the Kalman filter: forward scan. From above to below at $k = 90$: $\hat{c}_4()$, $\hat{c}_3()$, $\hat{c}_2()$, $\hat{c}_1()$ and $\hat{b}()$.

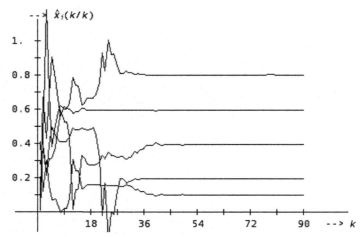

FIG. 6.6 The estimated state by the Kalman filter: backward scan (see Fig. 6.5).

designs (forward, backward, and optimal). It shows that the adapted measurement noise variance does not converge in one run. By processing through the measurements again in an iterative way, the right value for the measurement noise variance is found after a few runs (Fig. 6.10). Each run uses as starting values for adaptation the final values of the previous run. The adaptation by Eq. (6.27) is superior because it converges within one run close toward the correct value of the measurement noise variance. This is demonstrated in Figs. 6.11 and 6.12 for the various designs applied.

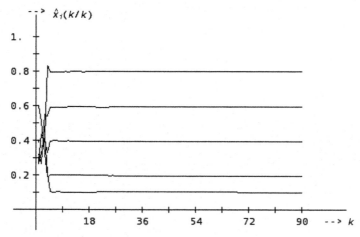

FIG. 6.7 The estimated state by the Kalman filter: optimal selection (see Fig. 6.5).

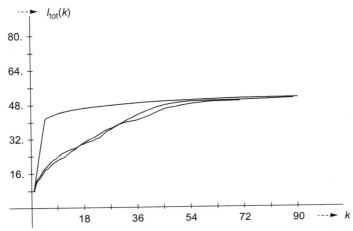

FIG. 6.8 The total information yields for various designs. *Upper curve* denotes optimal selection and *lower curves* forward and backward scan.

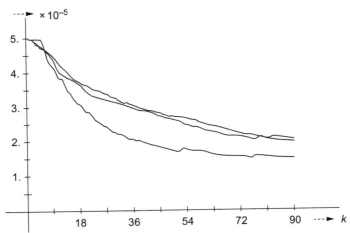

FIG. 6.9 Adaptation of the measurement noise variance $\hat{R}()$ by Eq. (6.25) for various designs (first run).

Example 6.2: The same individual molar spectra as in Example 6.1 (Fig. 6.3) are applied. For the state space model used, see Section 6.2.1. The multicomponent system comprises four components plus two extra terms for the stochastic drift. The concentrations are 0.2, 0.4, 0.6, and 0.8. The measurements corrupted by measurement noise and stochastic drift are depicted in Fig. 6.13. A measurement noise variance $R(k) = 10^{-5}$ and a nonzero diagonal element $Q_{66}(k-1) = 5.10^{-6}$ in the system noise covariance matrix $Q(k-1)$ are used.

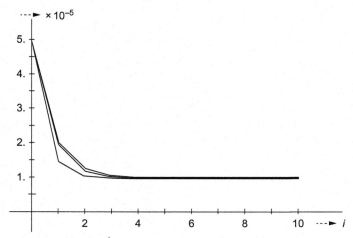

FIG. 6.10 Iterated adaptation $\hat{R}()$ by Eq. (6.25) recycling through the measurements for various designs.

Optimization cannot be done in this case because there is a stringent relation governed by the dynamics of the state space model applied. Also here, adaptation is employed in favor of restriction, which has been omitted in this example.

The estimated state by the Kalman filter scanning forward the measurements is plotted in Fig. 6.14. The simulated and estimated stochastic drift variable $d(k)$ or $x_5(k)$ are compared in Fig. 6.15. After about 50 measurements, the Kalman filter tracks the stochastic drift quite well and gives the concentrations their final

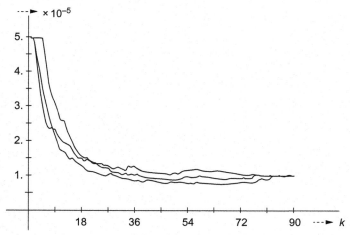

FIG. 6.11 Adaptation of the measurement noise variance $\hat{R}()$ by Eq. (6.27) for various designs (first run).

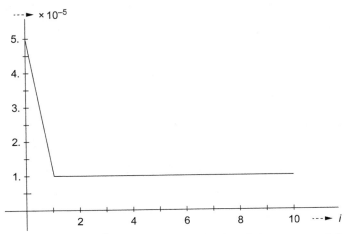

FIG. 6.12 Iterated adaptation $\hat{R}()$ by Eq. (6.27) recycling through the measurements for various designs.

values. The results by fixed-interval smoothing afterward are demonstrated in Figs. 6.16 and 6.17. It keeps the estimated concentrations constant backward in the sequence because there is no system noise involved on these unknowns in the state.

A look at the Figs. 6.14–6.17 reveals that stochastic drift may be viewed as the appearance of an unknown disturbance peak in a measured spectrum. If some impurity in the sample mixture would have a similar spectrum like the one for the applied stochastic drift, the Kalman filter and possibly fixed-interval

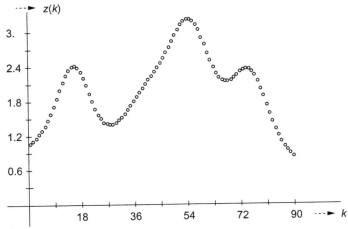

FIG. 6.13 Multicomponent system: the measurements corrupted by noise and stochastic drift.

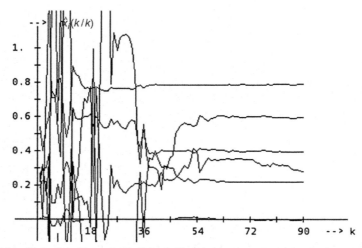

FIG. 6.14 The estimated state by the Kalman filter: forward scan. From above to below at $k = 90$: $\hat{c}_4()$, $\hat{c}_3()$, $\hat{c}_2()$, $\hat{d}()$, $\hat{c}_1()$ and $\hat{a}()$.

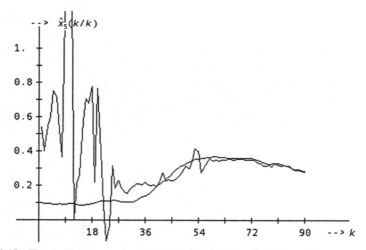

FIG. 6.15 The simulated and estimated stochastic drift $d()$ by the Kalman filter.

smoothing thereafter would do the job. Another source of stochastic drift is found when the pure component spectra deviate somewhat in the measured sample. Note that this approach cannot be done by classical least squares.

The adaptation of the system and measurement noise variances $\hat{Q}_{66}(k)$ and $\hat{R}(k)$ by Eqs. (6.42) and (6.45) are done starting from the initial values

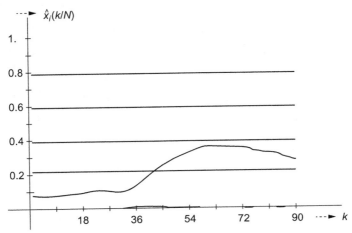

FIG. 6.16   The estimated state by fixed-interval smoothing (see Fig. 6.14).

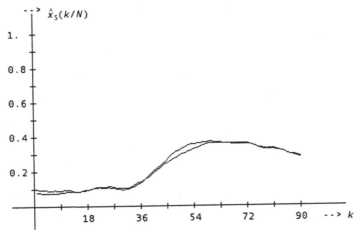

FIG. 6.17   The simulated and estimated stochastic drift $d()$ by fixed-interval smoothing.

$\hat{Q}_{66}(0) = 2.5 \cdot 10^{-5}$ and $\hat{R}(0) = 5 \cdot 10^{-5}$. Similar graphs for state estimation can be produced as before, which shall not be given. For plotting the system noise variance, $\hat{Q}_{66}(k)$ is multiplied with a factor 2. From Fig. 6.18 follows that the adapted system and measurement noise variances does not converge in one run. Again, an iterative procedure has to be applied to get close to their correct values, which is demonstrated in Fig. 6.19. It needs about 15 iterations to converge. Convergence of the measurement noise variance $\hat{R}(k)$ is fastest. Starting

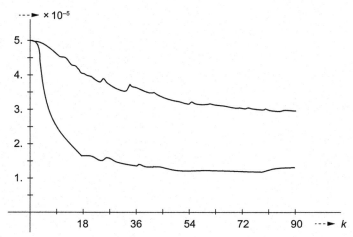

FIG. 6.18 Adaptation of the system and measurement noise variances (first run); *lower curve* $\hat{R}()$ and *upper curve* $2*\hat{Q}_{66}()$.

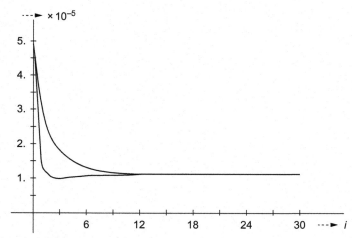

FIG. 6.19 Iterated adaptation recycling through the measurements (see Fig. 6.18).

from different initial values of $\hat{R}(0)$ and $\hat{Q}_{66}(0)$ leads to similar final results. In case there are two nonzero elements $\hat{Q}_{55}(k)$ and $\hat{Q}_{66}(k)$ in the system noise covariance matrix, the number of iterations increases to about 40. In practice, therefore, it is advised to keep the number of nonzero diagonal elements in the system noise covariance matrix as small as possible.

Chapter 7

The calibration system

Chapter outline

7.1 Linear calibration

7.1.1 Modeling

The performance of analytical systems can be assessed in terms of quality of results and times for which the system will function properly without breakdowns and/or disturbances. In general, the quality of systems can be resolved into two main components: A stationary random component associated with fluctuations under normal working conditions and a semirandom, nonstationary component caused by irreversible processes such as contamination, evaporation, wear, or aging. All analytical systems have some tendency toward nonstationary behavior. Varying systematic deviation or a time-dependent variance of the analytical errors can occur, or the system may even become defective, giving errors beyond some preset limit. Calibration drift in time is a particular problem in automated procedures. Drift can seriously affect the accuracy of the results for unknown samples if it is ignored.

As a start, one of the most frequent applications of estimation is probably the regression method for the determination of the sensitivities, i.e., the intercept a_0 and slope a_1 from data that are assumed to follow a linear relationship:

$$z(k) = a_1 c(k) + a_0 \qquad (7.1)$$

where $z(k)$ is the observed measurement corresponding to the kth sample with known concentration $c(k)$.

State Estimation in Chemometrics. https://doi.org/10.1016/B978-0-08-102603-8.00007-9

As a major source of error, in practical situations, the straightforward calibration model may be inadequate. Modifications can be made, for example with the terms $c^2(k)$, $1/c(k)$, $\log(c(k))$, or $10^{c(k)}$. In a similar way, the extension toward a higher-degree polynomial model is obtained. Hitherto, classical least squares still holds as the best estimation procedure.

A different approach has to be considered when drift is present in the linear calibration graph. Each parameter is extended in order to describe the time-dependent drifting behavior. A stochastic drifting parameter can be described by using two random variables. If stochastic drift acts on both the slope and intercept of a linear calibration graph, the state will expand with two extra parameters, which describes the slowly varying behavior of the system.

The state space model for a drifting linear calibration graph is as follows:

$$\begin{pmatrix} a_1(k) \\ a_0(k) \\ \alpha_1(k) \\ \alpha_0(k) \end{pmatrix} = \begin{pmatrix} 1 & 0 & 1 & 0 \\ 0 & 1 & 0 & 1 \\ 0 & 0 & 1 & 0 \\ 0 & 0 & 0 & 1 \end{pmatrix} \begin{pmatrix} a_1(k-1) \\ a_0(k-1) \\ \alpha_1(k-1) \\ \alpha_0(k-1) \end{pmatrix} + \mathbf{w}(k-1) \tag{7.2}$$

$$z(k) = (c(k) \quad 1 \quad 0 \quad 0) \begin{pmatrix} a_1(k) \\ a_0(k) \\ \alpha_1(k) \\ \alpha_0(k) \end{pmatrix} + v(k)$$

where $\alpha_0()$ and $\alpha_1()$ are the extended drift parameters to be estimated.

The observability matrix of the drifting linear calibration graph is given by:

$$M = \left(\mathbf{h} \quad F^{\mathrm{T}}\mathbf{h} \quad F^{\mathrm{T}}F^{\mathrm{T}}\mathbf{h} \quad F^{\mathrm{T}}F^{\mathrm{T}}F^{\mathrm{T}}\mathbf{h} \right) = \begin{pmatrix} c_1 & c_2 & c_3 & c_4 \\ 1 & 1 & 1 & 1 \\ 0 & c_2 & 2c_3 & 3c_4 \\ 0 & 1 & 2 & 3 \end{pmatrix} \tag{7.3}$$

The determinant of the matrix $|M|$ is zero if all the concentrations c_i are the same, have an increasing order 1,2,3,4 or in the reverse way 4,3,2,1, but the system is observable, i.e., $|M| \neq 0$ if there are two different concentrations, each measured twice.

7.1.2 Prediction

Prediction is done by Eqs. (3.11) given in Section 3.3. For each evaluation of an unknown sample, the skipping equations (3.12) in Section 3.3 are employed.

7.1.3 Filtering

Filtering is done by Eqs. (3.26) of the Kalman filter in Section 3.4.

7.1.4 Smoothing

Eqs. (3.28) in Section 3.6.1 for fixed-interval smoothing are of particular interest. These equations can be applied even in case the calibrations are on variable positions with inbetween unknowns in a sample batch.

7.1.5 Evaluation

In accordance with the dynamics involved in the drifting calibration model, the Kalman filter predicts the varying state, which is used to evaluate unknown samples. Because of the dynamics and system noise involved, the covariance matrix or the uncertainty in the estimated state is increased by prediction. When a measurement is processed for evaluation of an unknown sample, the predicted state and covariance matrix remain the same. This is done by using the simplified equations (3.12) for skipping filtering in Section 3.3.

The concentration $\hat{c}(k)$ of an unknown sample is obtained by rewriting the calibration function (7.1) introducing the lower and upper confidence limits $\underline{c}(k)$ and $\overline{c}(k)$:

$$\hat{c}(k) = \{z(k) - \hat{a}_0\}/\hat{a}_1 \tag{7.4}$$

$$\text{var}\{\hat{c}(k)\} = \frac{1}{\hat{a}_1^2}\{\mathbf{h}^T(k)P(k/k-1)\mathbf{h}(k) + R(k)\}$$

with: $\mathbf{h}^T(k) = (\hat{c}(k)\ \ 1\ \ 0\ \ 0)$ and $\hat{a}_0$, $\hat{a}_1$ taken from $\hat{\mathbf{x}}(k/k-1)$
and: $\underline{c}(k), \overline{c}(k) = \hat{c}(k) \pm t(1 - 1/2\alpha, \infty)\sqrt{\text{var}\{\hat{c}(k)\}}$
The variance in $\hat{c}(k)$ is based on a first-order error propagation:

$$y = f(x_1, x_2, \ldots, x_n) \tag{7.5}$$

$$\text{var}\{y\} \approx \sum_{i=1}^{n}\sum_{j=1}^{n}\left(\frac{df}{dx_i}\right)^T \text{cov}(x_i, x_j)\left(\frac{df}{dx_j}\right)$$

where df/dx_i is the first derivative of the function with respect to x_i and $\text{cov}(x_i, x_j)$ is the covariance between the parameters. Straightforward development of (7.5) applied to the first equation (7.4) gives the variance in the evaluated result.

The drift parameters are not directly involved in the variance formula; they affect it indirectly through the dynamics of the system.

A more general formula for the evaluated result can be derived from the predicted measurement with its variance:

$$E\{z(k)\} = \hat{z}(k) = \mathbf{h}^T(k)\hat{\mathbf{x}}(k/k-1) \tag{7.6}$$

$$E\left\{(z(k) - \hat{z}(k))(z(k) - \hat{z}(k))^T\right\} = \mathbf{h}^T(k)P(k/k-1)\mathbf{h}(k) + R(k)$$

One gets the evaluated result with its lower and upper confidence limits by solving the following equations given in state space notation:

$$\hat{c}(k) \Leftarrow z(k) - \mathbf{h}^{\mathrm{T}}(k)\hat{\mathbf{x}}(k/k-1) = 0 \tag{7.7}$$

$$\underline{c}(k), \overline{c}(k) \Leftarrow \{z(k) - \mathbf{h}^{\mathrm{T}}(k)\hat{\mathbf{x}}(k/k-1)\}^{2}$$

$$-F(1-\alpha, 1, \infty)\{\mathbf{h}^{\mathrm{T}}(k)P(k/k-1)\mathbf{h}(k) + R(k)\} = 0$$

Formulas (7.6) and (7.7) hold generally for any calibration graph with or without incorporation of stochastic drift. The presented equations (7.7) are iteratively solved by means of a bisection procedure, which is stopped when a preset precision is reached. Firstly, one computes the evaluated result for an unknown sample from the measurement. Thereafter, follows computation of both confidence limits; the bisection is initiated twice each time with the previously evaluated result as one of its starting values. In this way, one obtains the evaluated result with its lower and upper confidence limits. The confidence interval by Formula (7.7) is asymmetric around the evaluated result. If no solution exists, the bisection procedure iterates to very bad values. This might happen in the initial unobservable stage or when the local slope is near zero.

The critical value $F(1-\alpha, 1, \ell-n) = t^{2}(1-\alpha/2, \ell-n)$ has to be used in Eq. (7.7) for a given confidence level α and $(1, \ell-n)$ degrees of freedom. ℓ denotes the number of measurements used for calibration. In the special case of a priori known noise variances, the critical value simplifies to $F(1-\alpha, 1, \infty) = z^{2}(1-\alpha/2)$, i.e., for a 95% confidence level, the numerical value $3.84 (=1.96^{2})$ should be used.

For a higher-degree polynomial calibration graph, the evaluated result for an unknown sample is resolved by the bisection method as given in Eq. (7.7). Here, the local slope or the first derivative with respect to $c(k)$ can be used for the division term in the variance formula (7.4). Thus, for a quadratic calibration graph $z(k) = a_{2}c(k)^{2} + a_{1}c(k) + a_{0}$, one uses the division $\{2\hat{a}_{2}\hat{c}(k) + \hat{a}_{1}\}^{2}$ instead of $\hat{a}_{1}^{2}$ in accordance with an extended measurement vector $\mathbf{h}^{\mathrm{T}}(k)$. Comparison of Eqs. (7.4) and (7.7) shows that especially at the boundaries of the working range in a polynomial calibration graph, there are differences in the computed confidence limits.

A diagram for the evaluation of an unknown sample is depicted in Fig. 7.1.

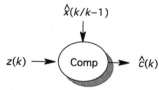

FIG. 7.1   Evaluation of an unknown sample.

In case of evaluation by fixed-interval smoothing the identities $\hat{\mathbf{x}}(k/N) \equiv \hat{\mathbf{x}}(k/k-1)$ and $P(k/N) \equiv P(k/k-1)$ in the given equations are used.

If an unknown sample is evaluated from more measurements in the sequence, the different concentration estimates are weighted by their variances.

7.1.6 Verification

Here, Eqs. (4.4) and (4.5) from Chapter 4 hold for the mean and chi-square value based on ℓ standardized innovations or residuals. One can compare the $N-\ell$ evaluated results with the known simulated values by the relation:

$$\chi^2(N-\ell) \approx 4t^2(1-\alpha/2, \infty) \sum_{i=1}^{N-\ell} \left(\frac{\hat{c}(i)-c(i)}{\overline{c}(i)-\underline{c}(i)} \right)^2 \tag{7.8}$$

In this way, it is possible to check the validity and correctness of the variance formulas (7.4) and (7.7) for the evaluated results.

7.1.7 Control

The maximal imprecision allowed for the expected evaluated results could be predefined as a proper analytical goal. The proposed design of quality control starts with the decision when to recalibrate followed by selection of the concentration standard that gives the best performance. The Kalman filter predicts in time the varying state of the calibration model, which is used for the evaluation of unknown samples. Because of the system noise and dynamics involved, the covariance matrix or the uncertainty in the estimated state is increased by prediction. The actual concentration to be analyzed is, of course, not known in advance. Therefore, all the available concentration standards are used for the computation. From a worst point of view, the maximum value N_{max} of the computed relative imprecisions $N\%$ is defined as a proper criterion:

$$\text{Maximize } N\% = \left\{ \frac{c(k)-\underline{c}(k)}{c(k)}, \frac{\overline{c}(k)-c(k)}{c(k)} \right\} *100\% \quad \text{for all } c(k) \tag{7.9}$$

$$N_{max} \geq N_{crit} \Rightarrow calibrate$$

$$N_{max} < N_{crit} \Rightarrow assessment$$

Consequently, a decision is made by comparing the maximum imprecision N_{max} with a preset critical imprecision N_{crit} to be chosen. The entire procedure under consideration is completed before any data on unknown samples are processed. The decision to do an assessment again enables the state vector and covariance matrix to be predicted one or more steps in advance until the critical imprecision is reached.

The greatest errors are found at the limiting boundaries of the working range in the calibration graph; the lowest concentration has the highest relative imprecision, so the control algorithm has to compute the criterion $N\%$ only once. It should be noted that the entire quality control procedure depends not only on the preset critical imprecision but also on the experimental character-istics of the state space model implemented, i.e., system and measurement noise variances, current estimated slope and concentration standards available for calibration.

The system noise variances can be interpreted as an important quality char-acteristic. In analytical practice, much effort is expended to minimize its effects in order to get a "*good*" system. However, if chemical, technical, or instrumen-tal improvements can only be achieved up to a certain limit, computational means are the only feasible way of further progress. The proposed algorithms describe the analytical possibilities completely. It shows that a "*bad*" system cannot acquire high precision. The results presented later demonstrate that one could make a reasonable choice for the critical imprecision level to be implemented.

7.1.8   Optimization

The entire data processing scheme for state estimation with prediction, filtering, and smoothing gives the information yield $I(k)$:

$$I(k) = I_{pre}(k) + I_{fil}(k) + I_{smo}(k) \tag{7.10}$$

$$I_{pre}(k) = 1/2ld\{|P(k-1/k-1)|/|F(k,k-1)P(k-1/k-1)F^{T}(k,k-1)+Q(k-1)|\}$$

$$I_{fil}(k) = 1/2ld\{\mathbf{h}^{T}(k)P(k/k-1)\mathbf{h}(k)/R(k)+1\}$$

$$I_{smo}(k) = 1/2ld\{|P(k+1/N)|/|P(k/N)|\}$$

The information $I(k)$ is given in two parts: online for the terms $I_{pre}(k)$ and $I_{fil}(k)$ used by the Kalman filter and offline $I_{smo}(k)$ for fixed-interval smoothing.

The system noise covariance matrix $Q(k-1)$ has a direct bearing on the magnitude of the covariance matrix $P(k/k-1)$. Less obvious is the effect of the transition matrix $F(k,k-1)$ on covariance behavior. The covariance matrix or the uncertainty in the estimated state is increased by the prediction equation. The corresponding loss of information is represented by $I_{pre}(k)$. Each time the state is not measured the information gain $I_{fil}(k)$ is zero and the total information yield $I_{tot}(k) = \sum_{i=1}^{k} I(i)$ decreases. The filtering equations reduce the uncer-tainty by means of the positive information gain $I_{fil}(k)$ supplied by the next calibration.

The optimal design problem is essentially a sequence of maximized exper-imental choices with the information yield as criterion. For an optimal perfor-mance, in general, the system and measurement noise variances $Q(k-1)$ and $R(k)$ should be minimized. More complicated are the effects of the elements in the transition matrix $F(k,k-1)$ and the measurement vector $\mathbf{h}(k)$.

The offline information yield $I_{smo}(k)$ for fixed-interval smoothing depends exclusively on the system noise covariance matrix $Q(k)$. If $Q(k)$ is the null matrix and $F(k+1,k)$ the identity matrix, the information yield $I_{smo}(k)$ diminishes to zero because $P(k/N)$ becomes equal to $P(k+1/N)$. However, the net effect for the information yield is balanced with respect to the minimization of the system noise covariance matrix $Q(k)$. Additionally, offline smoothing improves the information gained online by the Kalman filter.

It should be noted that the model structure used embodies in principle all intuitive rules of thumb on optimization in estimation problems. This will be demonstrated next for a drifting linear calibration graph.

In the drifting calibration system, the transition matrix $F(k,k-1)$ and the system noise covariance matrix $Q(k-1)$ are constant; the concentration $c(k)$ in the measurement vector $\mathbf{h}(k)$ is the only variable that can be manipulated experimentally. This variable is only involved in the filtering part but not in the prediction part of the information yield. Thus, only the information yield $I_{fil}(k)$ has to be maximized.

The decision on which concentration should be used to calibrate the system is given by:

$$\text{Maximize} \quad \mathbf{h}^T(k)P(k/k-1)\mathbf{h}(k)/R(k) \quad \text{for all } c(k) \qquad (7.11)$$

where the optimization is done for all the available concentration standards.

Or for a first degree or linear calibration graph:

$$\text{Maximize} \quad c^2(k)P_{11}/R(k) + 2c(k)P_{12}/R(k) + P_{22}/R(k) \qquad (7.12)$$

where P_{ij} is the i,jth element of the covariance matrix $P(k/k-1)$.

This equation has a minimum at $c(k) = -P_{12}/P_{11}$. Note that the element P_{12} is negative, which makes the computed minimum $c(k)$ positive lying somewhere within the working range. Because of the limited concentration range, the optimal position in the linear calibration graph is always found at one of the boundaries. The boundaries to be chosen in the sequence will be alternated as required for observability. The effect of the concentration range is difficult to show. However, if the optimal positions are found at the boundaries of the working range, there are still better concentrations outside this range. In summary:

- $I_{fil}(k)$ increases with a decreasing noise variance $R(k)$.
- $I_{fil}(k)$ increases if the concentration range increases.
- $I_{fil}(k)$ is maximal at one of the boundaries of the working range.

For a quadratic calibration graph in addition to the boundaries, also the center of the working range will be found as one of the optimal positions.

More generally, when a higher-degree polynomial calibration graph is used, the optimal positions chosen from a working range $[-1, +1]$ are given in

TABLE 7.1 Optimal positions in a polynomial calibration graph.

Degree	$c(k)$					
1	−1.0	+1.0				
2	−1.0	0.0	+1.0			
3	−1.0	−0.447	+0.447	+1.0		
4	−1.0	−0.655	0.0	+0.655	+1.0	
5	−1.0	−0.765	−0.285	+0.285	+0.765	+1.0

Table 7.1. Here, the offline optimal design computation for the determined solution is used with no system noise and no drift on the calibration graph.

For the imprecision of the evaluated results $\hat{c}(k)$, the corresponding information yield $I_{un}(k)$ follows from Eq. (7.4):

$$I_{un}(k) = 1/2ld\{\,\mathrm{var}(c_0)/\mathrm{var}(\hat{c}(k))\,\} \tag{7.13}$$

$$= 1/2ld\{\,\mathrm{var}(c_0)\hat{a}_1^2/\{\mathbf{h}^{\mathrm{T}}(k)P(k/k-1)\mathbf{h}(k)+R(k)\}\,\}$$

From Eq. (7.13), it follows that:

− $I_{un}(k)$ increases with an increasing slope $\hat{a}_1$.
− $I_{un}(k)$ increases with a decreasing noise variance $R(k)$.
− $I_{un}(k)$ has a maximum somewhere within the working range.

The maximum is located at the mean of the used concentration standards, if system noise and stochastic drift are not involved. Furthermore, the resulting variance will become smaller for each additional measurement used for evaluation of the unknown sample. This means that the information yield $I_{un}(k)$ concerning the unknown sample increases.

All these results differ only slightly from the traditional conclusions on optimal designs in calibration problems. For a higher-degree polynomial calibration graph, results can be obtained similarly.

7.1.9   Adaptation

In theory, the adaptation algorithms of both the system and measurement noise variances $Q(k-1)$ and $R(k)$ are not suitable for a system alternating between calibrations and evaluations on variable positions in a batch (Eqs. 6.42 and 6.45 in Section 6.2.6). In practice, therefore, one has to do a separate adaptation run where the batch of samples contains only calibrations. The thus-adapted noise variances are further assumed to remain the same between different batches of samples.

Consequently, to perform evaluation of samples, one has to operate the calibration system with a priori known noise variances implemented. It is still possible for the full system alternating between calibrations and evaluations to do verification of the calibration measurements. If there is a significant difference of the mean and/or chi-square value based on the standardized innovations or residuals, one could decide to do a full recalibration again for adaptation of the system and measurement variances.

7.2 Nonlinear calibration

7.2.1 Modeling

In analytical practice, sometimes, a nonlinear calibration graph is found with an exponential function, i.e.:

$$z(k) = g(\mathbf{x}(k), k) = x_\infty \left\{ 1 - e^{-k_i c(k)} \right\} + x_0 \tag{7.14}$$

where $z(k)$ is the kth measurement, x_∞ the limiting value at infinity, k_i the exponential constant, x_0 the baseline, and $c(k)$ the concentration.

In principle, this equation can be resolved with nonlinear curve fitting as shown earlier in Chapter 2. One could easily develop a state space model with all the parameters of the exponential calibration graph suffering from drift. In case only the baseline is subject to stochastic drift, the following identities have to be used in the state space model, i.e., the state vector, transition matrix, and measurement vector are:

$$\mathbf{x}^T(k) = (x_\infty(k) \quad k_i(k) \quad x_0(k) \quad \alpha(k)) \tag{7.15}$$

$$F(k, k-1) = \begin{pmatrix} 1 & 0 & 0 & 0 \\ 0 & 1 & 0 & 0 \\ 0 & 0 & 1 & 1 \\ 0 & 0 & 0 & 1 \end{pmatrix}$$

$$\mathbf{h}^T(\mathbf{x}(k), k) = \frac{dg(\mathbf{x}(k), k)}{d\mathbf{x}(k)}$$

$$= \left(1 - e^{-k_i(k)c(k)} \quad x_\infty(k)c(k)e^{-k_i(k)c(k)} \quad 1 \quad 0 \right)$$

where $\alpha(k)$ is the extended drift parameter. Note that here the applied state space model involves a linear system equation and a nonlinear measurement equation.

7.2.2 Prediction

Prediction is done by Eqs. (3.11) given in Section 3.3. For each evaluation of an unknown sample, the skipping equations (3.12) are used.

7.2.3 Filtering

Because only the measurement equation is nonlinear, this allows the use of the extended Kalman filter (Eqs. 5.5 in Section 5.2) or the iterated extended Kalman filter (Eqs. 5.6 and 5.8 in Section 5.3). In the last one, nonlinear filtering improves the estimated state $\hat{\mathbf{x}}^i(k/k)$ iteratively starting from $\hat{\mathbf{x}}(k/k-1)$.

7.2.4 Smoothing

Because the system equation is linear, no serious problems have to be expected in applying smoothing after all the measurements have been processed by the extended Kalman filter or the iterated extended Kalman filter. Eqs. (3.28) in Section 3.6.1 for fixed-interval smoothing are of particular interest.

7.2.5 Evaluation

Here, one has to fall back to the general equations in state space notation for the predicted measurement with its variance (7.6) and also the evaluated result with its lower and upper confidence limits (7.7):

$$E\{z(k)\} = \hat{z}(k) = g(\hat{\mathbf{x}}(k/k-1), k) \tag{7.16}$$

$$E\left\{(z(k) - \hat{z}(k))(z(k) - \hat{z}(k))^{\mathrm{T}}\right\} = \mathbf{h}^{\mathrm{T}}(\hat{\mathbf{x}}(k/k-1), k)P(k/k-1)\mathbf{h}(...) + R(k)$$

$$\hat{c}(k) \Leftarrow z(k) - g(\hat{\mathbf{x}}(k/k-1), k) = 0 \tag{7.17}$$

$$\underline{c}(k), \overline{c}(k) \Leftarrow \{z(k) - g(\hat{\mathbf{x}}(k/k-1), k)\}^2$$

$$-F(1-\alpha, 1, \infty)\left\{\mathbf{h}^{\mathrm{T}}(\hat{\mathbf{x}}(k/k-1), k)P(k/k-1)\mathbf{h}(...) + R(k)\right\} = 0$$

$$\mathbf{h}^{\mathrm{T}}(\hat{\mathbf{x}}(k/k-1), k) = \frac{dg(\mathbf{x}(k), k)}{d\mathbf{x}(k)}\bigg|_{\mathbf{x}(k) = \hat{\mathbf{x}}(k/k-1)}$$

Eqs. (7.17) are resolved by the bisection method as described before.

In case of evaluation by fixed-interval smoothing, the identities $\hat{\mathbf{x}}(k/N) \equiv \hat{\mathbf{x}}(k/k-1)$ and $P(k/N) \equiv P(k/k-1)$ in the given equations are used.

7.2.6 Verification

This is similar as described previously in Section 7.1.6.

7.2.7 Control

The procedure stays the same as described before in Section 7.1.7.

7.2.8 Optimization

In general, from the information yield $I(k)$ (Eq. 7.10), it follows that the system and measurement noise variances $Q(k-1)$ and $R(k)$ should be minimized for an optimal performance. See further Section 7.1.8. Here, optimization is based on maximizing the following equation in an online fashion:

$$\text{Maximize}\ \ \mathbf{h}^T(\hat{\mathbf{x}}(k/k-1),k)P(k/k-1)\mathbf{h}(\ldots)/R(k)\ \ \text{for all}\ c(k) \qquad (7.18)$$

$$\mathbf{h}^T(\hat{\mathbf{x}}(k/k-1),k)=\left.\frac{dg(\mathbf{x}(k),k)}{d\mathbf{x}(k)}\right|_{\mathbf{x}(k)=\hat{\mathbf{x}}(k/k-1)}$$

where the optimization is done for all the available concentration standards. Eq. (7.18) holds for either the extended Kalman filter or the iterated extended Kalman filter.

7.2.9 Adaptation

There is no reason to presume why the adaptation of the system and measurement noise variances in a nonlinear case would not work. When in the initial phase, the estimated state is too wrong; sometimes, too high noise variances are produced by the adaptation algorithms. We refer further to Section 7.1.9.

7.3 Examples

Example 7.1: For the applied state space model of a linear calibration graph involving stochastic drift, see Section 7.1.1. One could develop easily a higher-degree polynomial calibration graph extended with drift on each of the parameters. An example of a varying state for the calibration system is depicted in Fig. 7.2. The strongly varying states around zero $\alpha_1(k)$ and $\alpha_0(k)$ are multiplied by a factor 25. From here, the measurements corrupted by stochastic drift and measurement noise are generated in Fig. 7.3. The solid lines are measurements used for calibrations, while the dotted lines are used for the evaluation of unknowns. The same holds for the concentrations involved, which are plotted in Fig. 7.4. For calibration, the concentration standards 1, 2, 3, 4, and 5 are available.

The measurement noise variance is $R(k)=10^{-5}$ and the nonzero diagonal elements of the system noise covariance matrix $Q(k-1)$ are $Q_{33}(k-1)=Q_{44}(k-1)=10^{-8}$.

A critical imprecision $N_{crit}=10\%$ for quality control and optimal selection of the concentration standards is employed. The calibration system has the freedom to control its behavior completely, except at the end of the run when it is forced to recalibrate. The calibration system starts with four calibrations, where the highest and lowest concentration standards are repeated twice as required for observability. Hereafter, it alternates regularly between calibrations and

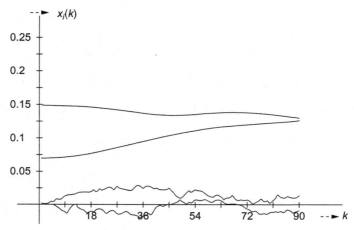

FIG. 7.2   Linear calibration system: the simulated state. From *above* to *below*: $a_1()$, $a_0()$, $25*\alpha_0()$, and $25*\alpha_1()$.

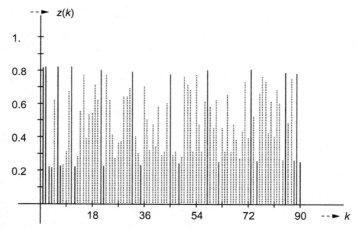

FIG. 7.3   The measurements corrupted by noise and stochastic drift. (—) used for calibration; (...) used for evaluation.

evaluations. Furthermore, for calibration, the highest and lowest concentration standards are alternated in the sequence as required for observability.

The estimated state by the Kalman filter is plotted in Fig. 7.5. The Kalman filter can track the real state quite well. Even better results are obtained by applying fixed-interval smoothing afterward. The concentrations involved are depicted in Fig. 7.6 and the estimated state by fixed-interval smoothing is shown in Fig. 7.7.

The evaluated relative errors, where the estimated concentrations are compared with their simulated values, can be found in histograms in Figs. 7.8 and 7.9 for the

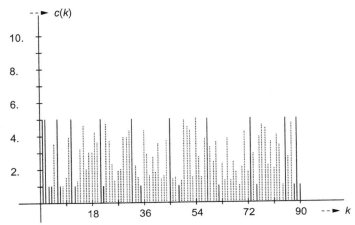

FIG. 7.4   The concentrations involved for the Kalman filter. (—) calibration values; (…) evaluated results.

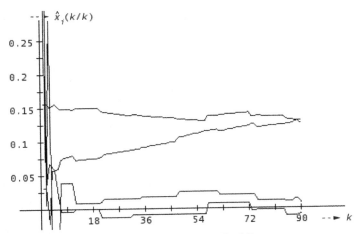

FIG. 7.5   The estimated state by the Kalman filter (see Fig. 7.2).

Kalman filter and fixed-interval smoothing, respectively. It shows that all the evaluated results are below the applied critical imprecision of 10%. A variance reduction of 66% is obtained from the formula $VR = (1 - s_{sm}^2/s_{kf}^2) \cdot 100\%$ when the results of the Kalman filter are compared with fixed-interval smoothing. If a time delay is not detrimental, fixed-interval smoothing should be used.

The maximum imprecisions computed are compared in Fig. 7.10 with the preset critical imprecision. The zigzag curve is generated by the Kalman filter, while the lower almost straight line follows from fixed-interval smoothing. For the applied calibration system, it is impossible for the Kalman filter to get a final

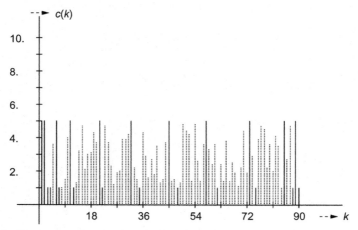

FIG. 7.6   The concentrations involved for fixed-interval smoothing (see Fig. 7.4).

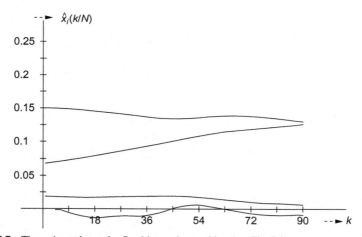

FIG. 7.7   The estimated state by fixed-interval smoothing (see Fig. 7.2).

imprecision lower than 6%, which is reduced by fixed-interval smoothing to about 5%. To get these imprecisions, all measurements should become calibrations.

The information yields of the Kalman filter and fixed-interval smoothing are plotted in Fig. 7.11. It shows clearly the positive spikes when the Kalman filter recalibrates. Here, also the total information yields divided by a factor 10 are depicted. The total information yield converges to a constant level and is improved by fixed-interval smoothing.

For the adaptation of the system and measurement noise variances $\hat{Q}_{33}(k)$, $\hat{Q}_{44}(k)$, and $\hat{R}(k)$, all measurements should be calibrations starting from

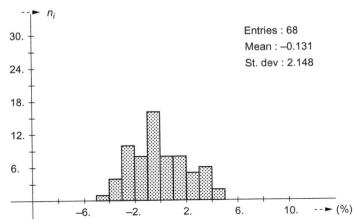

FIG. 7.8 The relative errors for evaluation by the Kalman filter.

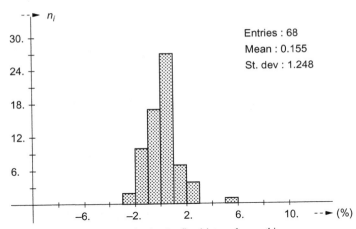

FIG. 7.9 The relative errors for evaluation by fixed-interval smoothing.

$\hat{Q}_{33}(0) = \hat{Q}_{44}(0) = 5 \cdot 10^{-8}$ and $\hat{R}(0) = 5 \cdot 10^{-5}$. The same drifting system as in Fig. 7.2 is employed. The results of the first run through the measurements are shown in Fig. 7.12. For plotting, the system noise variances are multiplied by a factor 1000. It demonstrates that the adapted noise variances are not converged to their right values in one run. In Fig. 7.13, an iterative procedure was applied to get close to the correct values after about 80 cycles through the measurements. Convergence of the measurement noise variance $\hat{R}(k)$ is fastest.

Example 7.2: For the state space model of a nonlinear calibration function that includes stochastic drift, see Section 7.2.1. An example of the varying state for a nonlinear calibration system extended with one drift term is depicted in Fig. 7.14. For plotting $x_{\infty}(k)$ is divided by a factor 4 and the strongly varying

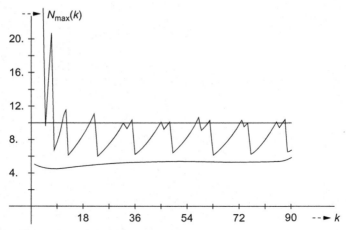

FIG. 7.10   The maximum imprecisions compared with the critical value. *Zigzag curve*: Kalman filter; *lower curve*: fixed-interval smoothing.

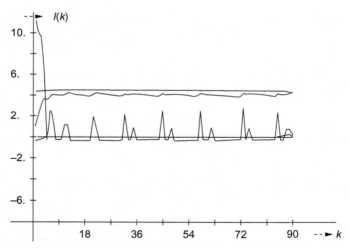

FIG. 7.11   The information yields (*lower curve*) of the Kalman filter and fixed-interval smoothing. *Upper curve*: total information yields/10.

drift variable $\alpha(k)$ is multiplied with a factor 5. One could develop easily a state space model with all the parameters of the exponential calibration graph suffering from stochastic drift. The measurements corrupted by stochastic drift and measurement noise are generated in Fig. 7.15. The solid lines are measurements used for calibrations, while the dotted lines are used for the evaluation of unknowns. The same holds for the concentrations involved, which are plotted

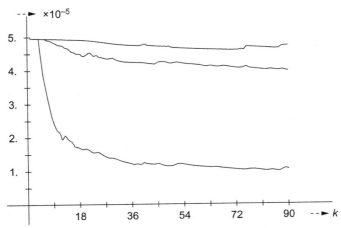

FIG. 7.12 Adaptation of the system and measurement noise variances (first run); *lower curve* $\hat{R}()$, *middle* $1000*\hat{Q}_{33}()$, and *upper curve* $1000*\hat{Q}_{44}()$.

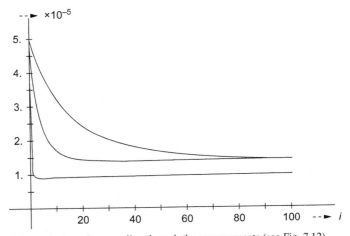

FIG. 7.13 Iterated adaptation recycling through the measurements (see Fig. 7.12).

in Fig. 7.16. For calibration, the concentration standards 1, 2, 3, 4, and 5 are available. A measurement noise variance $R(k)=10^{-5}$ and a nonzero diagonal element $Q_{44}(k-1)=10^{-6}$ in the system noise covariance matrix $Q(k-1)$ are employed. A critical imprecision $N_{crit}=10\%$ is used for quality control. Serious problems are encountered with optimization in this nonlinear case because sometimes-unobservable designs are produced. Further application of optimal selection has been omitted. To improve the convergence of nonlinear

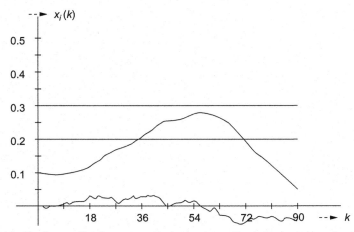

FIG. 7.14   Nonlinear calibration system: the simulated state. From *above* to *below* at $k = 1$: $x_\infty()/4$, $k_i()$, $x_0()$, and $5*\alpha()$.

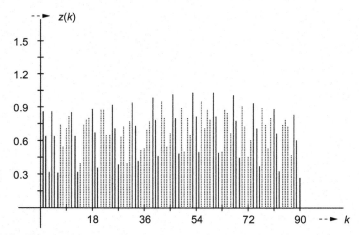

FIG. 7.15   The measurements corrupted by noise and stochastic drift. (—) used for calibration; (...) used for evaluation.

estimation, somewhat more calibrations are used as required for observability. After the initial start of six calibrations, the system alternates regularly between calibrations and evaluations. At the end of the run, the system is forced to recalibrate. The estimated state by the iterated extended Kalman filter operating with $\ell = 5$ is plotted in Fig. 7.17. The iterated extended Kalman filter follows the real state quite well. For the extended Kalman filter with $\ell = 1$, somewhat poorer results are found.

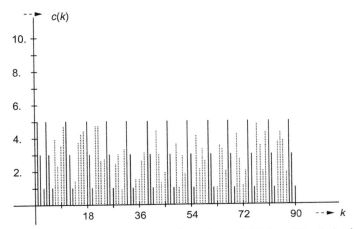

FIG. 7.16 The concentrations involved for the iterated extended Kalman filter. (—) calibration values; (...) evaluated results.

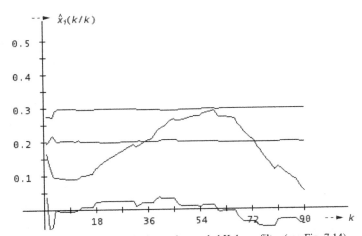

FIG. 7.17 The estimated state by the iterated extended Kalman filter (see Fig. 7.14).

If a time delay is not detrimental, improved results are obtained by applying fixed-interval smoothing afterward. The concentrations involved are shown in Fig. 7.18 and the estimated state by smoothing in Fig. 7.19.

The evaluated relative errors can be found in histograms in Figs. 7.20 and 7.21 for the Kalman filter and fixed-interval smoothing, respectively. It shows that all the evaluated results are below the employed critical imprecision of 10%. A variance reduction of 75% is obtained when the results of the iterated

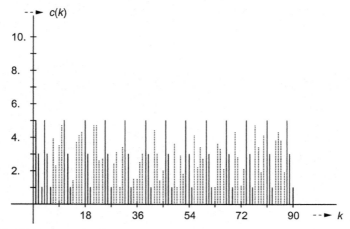

FIG. 7.18   The concentrations involved for fixed-interval smoothing (see Fig. 7.16).

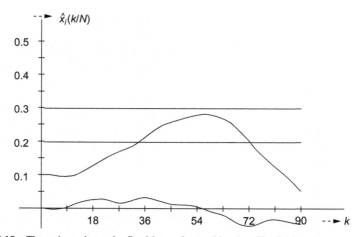

FIG. 7.19   The estimated state by fixed-interval smoothing (see Fig. 7.14).

extended Kalman filter are compared with the results for fixed-interval smoothing (see Example 7.1).

The maximum imprecisions computed are compared in Fig. 7.22 with the preset critical imprecision. The zigzag curve is generated by the Kalman filter, while the lower almost straight line follows from fixed-interval smoothing. For the applied calibration system, it is impossible for the Kalman filter to get a final imprecision lower than 5%, which is reduced by fixed-interval smoothing to

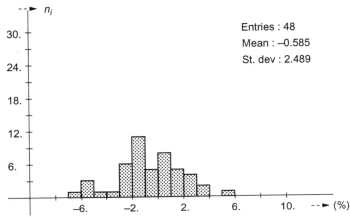

FIG. 7.20 The relative errors for evaluation by the iterated extended Kalman filter.

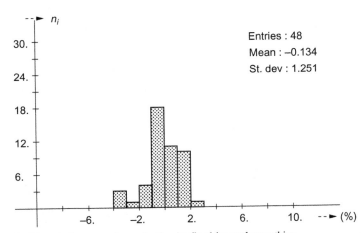

FIG. 7.21 The relative errors for evaluation by fixed-interval smoothing.

about 3.5%. To get these imprecisions, all measurements should become calibrations.

The information yields of the Kalman filter and fixed-interval smoothing are shown in Fig. 7.23. It shows positive spikes when the iterated extended Kalman filter recalibrates. Here, also the total information yields divided by a factor 10

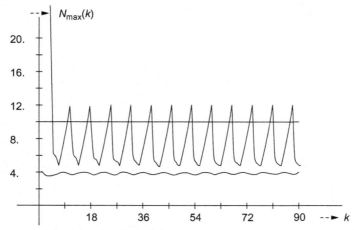

FIG. 7.22   The maximum imprecisions compared with the critical value. *Zigzag curve*: iterated extended Kalman filter; *lower curve*: fixed-interval smoothing.

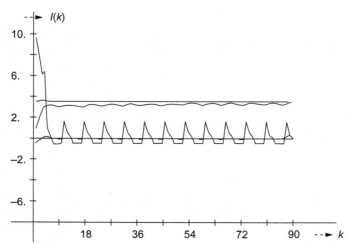

FIG. 7.23   The information yields (*lower curve*) of the iterated extended Kalman filter and fixed-interval smoothing. *Upper curve*: total information yields/10.

are depicted. The total information yield converges to a constant level and is improved by fixed-interval smoothing.

For the adaptation of the system and measurement noise variances $\hat{Q}_{44}(k)$ and $\hat{R}(k)$, all measurements are calibrations starting from $\hat{Q}_{44}(0) = 5 \cdot 10^{-6}$ and $\hat{R}(0) = 5 \cdot 10^{-5}$. Here, the same drifting system as in Fig. 7.14 is employed. In Fig. 7.24, the results for the first run of adaptation through the measurements

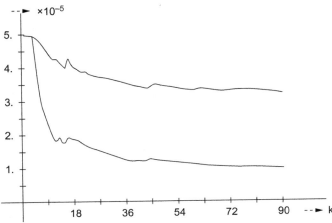

FIG. 7.24 Adaptation of the system and measurement noise variances (first run); *lower curve* $\hat{R}()$ and *upper curve* $10*\hat{Q}_{44}()$.

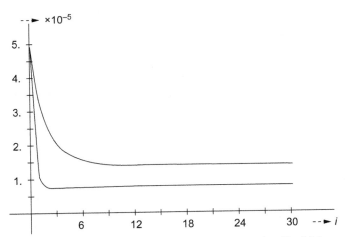

FIG. 7.25 Iterated adaptation recycling through the measurements (see Fig. 7.24).

are shown. For plotting purposes, $\hat{Q}_{44}(k)$ is multiplied by a factor 10. It demonstrates that the adapted system and measurement noise variances are not converged to their right values in the first run. In Fig. 7.25, an iterative procedure is applied to get close to its correct values after about 15 cycles through the measurements. Again, convergence of the measurement noise variance $\hat{R}(k)$ is fastest.

Chapter 8

The titration system

Chapter outline

8.1 Discrete titration

8.1.1 Modeling

In titrimetry, unknown samples are evaluated by adding titrant of a known concentration until an equivalence point corresponding to a stoichiometric chemical reaction or an empirically established endpoint has been reached. The equivalence point (or setpoint) is indicated by the change of a suitable physical property such as color, absorbance, potential, pH, current, or conductance. Its ease of operation, accuracy, and precision make the titration a widely used analytical method. Many titrators now utilize computers for control purposes, signal registration, data processing, and report generation. In practice, it is often assumed that in potentiometric titrations, the equivalence and inflection point coincide. Inflection point(s) are rapidly evaluated by utilizing the derivatives of the titration curve.

In automatic titrators, the continuous addition of titrant can lead to systematic error if the chemical reaction or electrode response involved is slow. If the titrant is added stepwise and equilibrium potentials are measured, accuracy is achieved. To obtain maximal precision within minimal analysis time, the online volume control for equidistant measurements is often preferred.

A nonlinear potentiometric titration curve is obtained by plotting the measured potentials versus the volume added and is more or less sigmoidal with one

State Estimation in Chemometrics. https://doi.org/10.1016/B978-0-08-102603-8.00008-0

or more inflections. If the titration curve is divided into infinitely small volume steps ∂V, a Taylor approximation for the potential x is allowed at volume V:

$$x(V + \partial V) = x(V) + \dot{x}(V)\partial V + 1/2\ddot{x}(V)\partial V^2 + \cdots \qquad (8.1)$$

with $\dot{x}()$ as the first derivative and $\ddot{x}()$ the second derivative. After truncation, a discrete version of a second-order expansion can be obtained. With successive steps from volume $V(k-1)$ to $V(k)$ in the titration curve and definitions of the identities $\Delta V \equiv V(k) - V(k-1)$, $k-1 \equiv V(k-1)$ and $k \equiv V(k)$, give the following set of difference equations:

$$x(k) = x(k-1) + \dot{x}(k-1)\Delta V + 1/2\ddot{x}(k-1)\Delta V^2 \qquad (8.2)$$

$$\dot{x}(k) = \dot{x}(k-1) + \ddot{x}(k-1)\Delta V$$

$$\ddot{x}(k) = \ddot{x}(k-1)$$

The given set of difference equations is a bad approximation of the nonlinear titration curve. There is also noise involved in the measurements.

Introduction of system and measurement noise with definition of $x_0() \equiv x()$, $x_1() \equiv \dot{x}()\Delta V$, $x_2() \equiv \ddot{x}()\Delta V^2$ give the required state space model:

$$\begin{pmatrix} x_0(k) \\ x_1(k) \\ x_2(k) \end{pmatrix} = \begin{pmatrix} 1 & 1 & 1/2 \\ 0 & 1 & 1 \\ 0 & 0 & 1 \end{pmatrix} \begin{pmatrix} x_0(k-1) \\ x_1(k-1) \\ x_2(k-1) \end{pmatrix} + \mathbf{w}(k-1) \qquad (8.3)$$

$$z(k) = (1 \quad 0 \quad 0) \begin{pmatrix} x_0(k) \\ x_1(k) \\ x_2(k) \end{pmatrix} + v(k)$$

The volume step ΔV is incorporated in the state $\mathbf{x}(k)$. In order to obtain the desired derivatives, the following similarity transformation should be applied:

$$\mathbf{x}(k) = \begin{pmatrix} x_0(k) \\ x_1(k) \\ x_2(k) \end{pmatrix} \qquad \mathbf{x}^*(k) = \begin{pmatrix} x(k) \\ \dot{x}(k) \\ \ddot{x}(k) \end{pmatrix} \qquad T(k) = \begin{pmatrix} 1 & 0 & 0 \\ 0 & \Delta V & 0 \\ 0 & 0 & \Delta V^2 \end{pmatrix} \qquad (8.4)$$

In this way, the state estimated by model (8.3) can be transformed for plotting purposes.

The observability criterion of the state space model (8.3) is satisfied, i.e., $|M| \neq 0$:

$$M = (\mathbf{h} \quad F^T\mathbf{h} \quad F^T F^T\mathbf{h}) = \begin{pmatrix} 1 & 1 & 1 \\ 0 & 1 & 2 \\ 0 & 1/2 & 2 \end{pmatrix} \qquad (8.5)$$

In the alternative state space model, the volume step ΔV is involved in the transition matrix and affects the observability criterion, which will converge to zero for $\Delta V \to 0$.

By extending the state space model to n terms, the transition matrix $F(k, k-1)$ will expand with the element $1/(n-1)!$ and the similarity matrix $T(k)$ with ΔV^{n-1}. Upon increasing the dimension n, the observability criterion $|M|$ will converge to zero, which limits the number of derivatives in the state.

As a result of the truncation of the discrete Taylor expansion, the last state variable stays in front a systematic error of $\Delta V/(n-1)!$ volume units.

It should be noted that the system noise exploits additional modeling errors to describe the nonlinearity of the titration curve caused by physical and chemical laws, i.e., conservation laws for mass and charge, equilibrium relations, Nernst equation. The input of the system is a certain volume addition of titrant (the sequence dependent variable) and the output is measured as a potential or pH.

The Kalman filter and fixed-interval smoothing can be directly applied to the given state space model (8.3). In practice, the measurement noise variance $R(k)$ to be used is known reasonable well in advance and can be found from the fluctuations in the measured potentials. The system noise covariance matrix $Q(k-1)$, especially the third diagonal element $Q_{33}(k-1)$, is used as a tuning parameter to describe the titration curve in a statistically correct way. $Q_{33}(k-1)$ describes the overall effects of the physical and chemical relations and quantities involved on the steepness of a particular inflection and is strongly dependent on the volume step ΔV.

It should be noted that the transition matrix and measurement vector in the state space model contain only constants. Thus, the covariance matrices in the Kalman filter and fixed-interval smoothing converge to a steady state. This means that the steady-state covariance matrices can be precomputed with inclusion of the associated gains. The Kalman filter and fixed-interval smoothing simplify into only processing the state equations for estimation, with a considerable improvement of the computation time and memory requirements. At this point, the estimation algorithms may be compared with the weighted digital filters based on polynomial least squares.

One alternative state space model should be mentioned. Applying the similarity transformation on the given state space model (8.3):

$$T(k) = \begin{pmatrix} 1 & k & k^2 \\ 0 & 1 & 2k \\ 0 & 0 & 2 \end{pmatrix} \quad \text{and:} \quad T^{-1}(k) = \begin{pmatrix} 1 & -k & 1/2k^2 \\ 0 & 1 & -k \\ 0 & 0 & 1/2 \end{pmatrix} \quad (8.6)$$

$$\text{gives:} \quad \mathbf{h}^{*\mathrm{T}}(k) = \mathbf{h}^{\mathrm{T}}(k)T(k) = \begin{pmatrix} 1 & k & k^2 \end{pmatrix}$$

$$F^*(k, k-1) = T^{-1}(k)F(k, k-1)T(k-1) = \begin{pmatrix} 1 & 0 & 0 \\ 0 & 1 & 0 \\ 0 & 0 & 1 \end{pmatrix}$$

Thus, in state space, the second-order derivative model is similar to a quadratic drift model.

8.1.2 Prediction

Prediction is done by Eqs. (3.11) given in Section 3.3. In case of nonuniformly spaced measurements resulting from variable volume additions, the skipping equations (3.12) in Section 3.3 are employed.

8.1.3 Filtering

Filtering is assigned to Eqs. (3.26) of the Kalman filter in Section 3.4.

8.1.4 Smoothing

Fixed-interval smoothing by Eqs. (3.28) in Section 3.6.1 is employed afterward. When the titration is done with online control (see Section 8.1.7), we have to deal with nonequidistant positions of the measurements in the sequence-dependent volume. One skips unnecessary predictions that are not needed to perform fixed-interval smoothing. The estimates $\hat{x}(k+i-1/k+i-1)$ and $\hat{x}(k+i-1/k-1)$ and covariance matrices $P(k+i-1/k+i-1)$ and $P(k+i-1/k-1)$ are stored to perform fixed-interval smoothing afterward. The corresponding transition matrix $F(k+i-1,k-1)$ or the value of i to store is given by:

$$F(k+i-1,k-1) = \begin{pmatrix} 1 & i & 1/2i^2 \\ 0 & 1 & i \\ 0 & 0 & 1 \end{pmatrix} \qquad (8.7)$$

This saves a lot of computation time and memory requirements.

8.1.5 Evaluation

Evaluation can only be done offline after processing all the measurements by the Kalman filter and fixed-interval smoothing. The inflection point in a potentiometric titration can be found from the first or second derivative of the estimated curve. Simple arithmetics determine the volume at which the second derivative is zero. Firstly, between given boundaries, the maximum and minimum of the second derivative are located. Secondly, within these extremes, a zero crossing of the second derivative occurs between the volumes $V(i-1)$ and $V(i)$. Hereafter, the inflection point V_{inf} in the titration curve is evaluated by linear interpolation:

$$V_{inf} = V(i-1) + \{\hat{x}_2(i-1/i-1)/[\hat{x}_2(i-1/i-1) - \hat{x}_2(i/i)]\}\Delta V \qquad (8.8)$$

$$V_{set} = V(i-1) + \{[x_{set} - \hat{x}_0(i-1/i-1)]/[\hat{x}_0(i/i) - \hat{x}_0(i-1/i-1)]\}\Delta V$$

where $\Delta V = V(i) - V(i-1)$, $\hat{x}_0()$ and $\hat{x}_2()$ are taken from the estimate $\hat{x}()$.

The setpoint volume V_{set} at a given potential x_{set} is obtained in a similar fashion. The titration curve may be either upward or downward. For fixed-interval smoothing, use in Eqs. (8.8) and (8.9) the identities

$\hat{x}_0(i-1/N) \equiv \hat{x}_0 \ (i-1/i-1), \ \hat{x}_0(i/N) \equiv \hat{x}_0(i/i), \ \hat{x}_2(i-1/N) \equiv \hat{x}_2(i-1/i-1),$ and $\hat{x}_2(i/N) \equiv \hat{x}_2(i/i).$

The variance in the evaluated results can be determined by using a first-order error propagation. Straightforward development of (7.5) in Section 7.1.5 applied to Eq. (8.8) gives for the variance in the evaluated inflection point:

$$\text{var}\{V_{inf}\} = \left\{ [\hat{x}_2^2(i-1/i-1) + \hat{x}_2^2(i/i)] / [\hat{x}_2(i-1/i-1) - \hat{x}_2(i/i)]^4 \right\} P_{33} \Delta V^2$$

(8.9)

where P_{33} denotes the third diagonal element of the covariance matrix $P(i/i)$ or $P(i/N)$. This equation only holds for equidistant volume additions assuming a steady-state P_{33}. Modification in case of variable volume additions is self-evident.

Similarly, a variance formula can be derived for the evaluated setpoint volume.

8.1.6   Optimization

From Eq. (7.10), it follows that for an optimal performance of the information yield $I(k)$ the system and measurement noise variances $Q(k-1)$ and $R(k)$ should be minimized, because the transition matrix $F(k,k-1)$ and measurement vector $\mathbf{h}(k)$ contain only constants. If the system noise covariance matrix $Q(k-1)$ is not the null matrix, the covariance matrix $P(k/k)$ by the Kalman filter approaches a steady state where an equilibrium is achieved between the information loss $I_{pre}(k)$ and information gain $I_{fil}(k)$. The total information yield $I_{tot}(k)$ will converge to a constant level. Finally, no more information $I(k)$ can be obtained by the Kalman filter and it operates following the changing state properly. The covariance matrix $P(k/N)$ by fixed-interval smoothing converges also to a steady state. After an initial period, the information gain $I_{smo}(k)$ diminishes to zero. Then, fixed-interval smoothing will still improve the predicted and filtered state.

The only way to influence the progress of the titration experimentally is to control the amount of volume addition. In order to get improved results, some general rules for optimization can be obtained from the variance formula (8.9) for the evaluated inflection point. The first term on the left side of Eq. (8.9) between curly brackets follows from the estimated state and can be treated as a constant. It will be further referred to as β. The corresponding information yield is:

$$I_{inf} = 1/2ld\{ \text{var}(V_0)/\text{var}(V_{inf}) \} = 1/2ld\{ \text{var}(V_0)/(\beta P_{33} \Delta V^2) \}$$ (8.10)

This implies that the acquired information will increase if either ΔV or P_{33} decrease. The system noise covariance matrix $Q(k-1)$ depends strongly on ΔV, i.e., if ΔV decreases, the elements of $Q(k-1)$ will also decrease. In addition, the value of P_{33} is related to the noise variances $Q(k-1)$ and $R(k)$, i.e., if either of them decreases, so will P_{33}. In summary, I_{inf} increases if either the volume step ΔV, system noise covariance matrix $Q(k-1)$, or measurement noise variance $R(k)$ decrease. For the evaluated setpoint, similar conclusions can be derived.

The information formula reflects the well-established rules of thumb that a minimal imprecision is found for the lowest possible ΔV and/or $R(k)$ values and that as many measurements as possible should be acquired around the inflection point(s). The constraints on imprecision are given by experimental characteristics, i.e., the discrimination of the titrant addition and the noise variance of the detector. Obviously, the estimation procedure can affect the accuracy of the evaluated result. If the titration is controlled online to obtain equidistant measurements using variable volume additions, very good imprecision with minimal analysis time may be approached as a proper analytical goal. For variable volume additions, the steady state of the covariance matrices is irrelevant.

8.1.7 Control

Online control in the titration to obtain equidistant measurements using variable volume additions can be done by a multistep prediction in the Kalman filter:

$$|\hat{x}_0(k+i-1/k-1) - \hat{x}_0(k-1/k-1)| \geq \Delta x_0 \qquad (8.11)$$

where "| |" denotes the absolute value. The titration curve may be either upward or downward. If the criterion Δx_0 is reached by predicting the titration curve $(i+1)$ steps forward, i volume units are added to the solution. In practice, minimal and maximal volume steps (ΔV_{min} and ΔV_{max}) are fixed in advance. The first two titrant additions are done with a volume step ΔV_{init} (i.e., three measurements) in order to provide an observable estimate. For variable volume additions, the steady-state approach of the Kalman filter and fixed-interval smoothing cannot be used.

8.1.8 Adaptation

Can only be done afterward in case of equidistant volume additions. Here, full adaptation of the noise variances gives bad results, due to the fact that there is a strongly varying system noise. Far from the inflection point(s), it is low and nearby inflection it becomes very high. This affects during the titration the progress of offline adaptation badly for both the system and measurement noise variances. The measurement noise variance is known reasonable well. Then, the adaptation should be limited to only processing the equations for the system noise variances (6.45) and skipping the equations for the measurement noise variance (6.42).

8.2 Continuous titration

8.2.1 Modeling

In general, automated titrations can be of two types, one in which the complete titration curve is monitored and the other in which the addition of titrant is stopped as soon as a preset signal is reached. In potentiometry, the endpoint titration with feedback control is based on the assumption that the overall response time of the reaction, mixing and the detector is fast. Obviously, the electrode must be reproducible and stable in the region of the established set-point. These conditions are fulfilled in many titrations, but if a slow response time is combined with a steep inflection, endpoint control may become time consuming and the results inaccurate because of overshoot. Correctly, these titrations should be done by the equilibrium method; after each addition of titrant, the signal is allowed to equilibrate before measurement. This last approach was described previously for discrete titrations.

This paragraph describes the extension of the state space model for continuous titrations in potentiometry. The titrant is added continuously and the variation in the signal is recorded simultaneously. To achieve the final results, the dynamics of the titration system have to be investigated.

A first-order linear dynamic system can be described by the following differential equation in the time domain:

$$\dot{x} + 1/T_x x = x_\infty \tag{8.12}$$

where x denotes the response at time t, $\dot{x}$ the first derivative, T_x is the time constant, and x_∞ is the steady state reached at $t = \infty$. For a sampling time Δt and the exponential factor $\Phi = e^{-\Delta t/T_x}$, the system can be represented in discrete terms by:

$$x(t + \Delta t) = \Phi x(t) + (1 - \Phi)x_\infty \tag{8.13}$$

Approximating the nonlinear titration curve in the volume domain with a second-order Taylor expansion readily provides the equations as given before:

$$x(V + \Delta V) = x(V) + \dot{x}(V)\Delta V + 1/2\ddot{x}(V)\Delta V^2 \tag{8.14}$$

$$\dot{x}(V + \Delta V) = \dot{x}(V) + \ddot{x}(V)\Delta V$$

$$\ddot{x}(V + \Delta V) = \ddot{x}(V)$$

where $x()$ denotes the function value, $\dot{x}()$ the first derivative, and $\ddot{x}()$ the second derivative at a given point V in the titration curve.

Replacing the term x_∞ in Eq. (8.13) by $x(V)$ and definition of the dynamic part $x_d(k) \equiv x(t + \Delta t)$, equilibrium parts $x_0(k) \equiv x(V + \Delta V)$, $x_1(k) \equiv \dot{x}(V + \Delta V)\Delta V$, and $x_2(k) \equiv \ddot{x}(V + \Delta V)\Delta V^2$ result in the required state space model:

$$\begin{pmatrix} x_d(k) \\ x_0(k) \\ x_1(k) \\ x_2(k) \end{pmatrix} = \begin{pmatrix} \Phi & 1-\Phi & 0 & 0 \\ 0 & 1 & 1 & 1/2 \\ 0 & 0 & 1 & 1 \\ 0 & 0 & 0 & 1 \end{pmatrix} \begin{pmatrix} x_d(k-1) \\ x_0(k-1) \\ x_1(k-1) \\ x_2(k-1) \end{pmatrix} + \mathbf{w}(k-1) \qquad (8.15)$$

$$z(k) = (1 \ 0 \ 0 \ 0) \begin{pmatrix} x_d(k) \\ x_0(k) \\ x_1(k) \\ x_2(k) \end{pmatrix} + v(k)$$

Volume and time units are related to each other by the equation $\Delta V = \rho \Delta t$, where ρ is the volume addition speed. The given set of difference equations is an inferior approximation of the nonlinear titration curve. In order to correct for modeling errors the system noise $\mathbf{w}(k-1)$ is introduced. Additional noise $v(k)$ is encountered on the measurements $z(k)$. If the continuous titration is monitored by potentiometry, the measurements $z(k)$ denote the measured potentials or pH's. The integration of the difference equations in time and volume introduces a systematic delay of one volume unit ΔV in the equilibrium part $x_0(k)$, $x_1(k)$, and $x_2(k)$ of the state. By means of a similarity transformation, an alternative state space model can be derived:

$$T(k) = \begin{pmatrix} 1 & 0 & 0 & 0 \\ 0 & 1 & 1 & 1/2 \\ 0 & 0 & 1 & 1 \\ 0 & 0 & 0 & 1 \end{pmatrix} \quad \text{gives:}$$

$$F^*(k, k-1) = \begin{pmatrix} \Phi & 1-\Phi & 1-\Phi & 1/2(1-\Phi) \\ 0 & 1 & 1 & 1/2 \\ 0 & 0 & 1 & 1 \\ 0 & 0 & 0 & 1 \end{pmatrix}$$

$$\text{and:} \quad \mathbf{h}^{*\mathrm{T}}(k) = (1 \ 0 \ 0 \ 0) \qquad (8.16)$$

The derived state space model suffers only a truncation error $1/2 \Delta V$ in the state $x_2(k)$. The observability criterion $|M| = |\mathbf{h} \ F^{\mathrm{T}}\mathbf{h} \ F^{\mathrm{T}}F^{\mathrm{T}}\mathbf{h} \ F^{\mathrm{T}}F^{\mathrm{T}}F^{\mathrm{T}}\mathbf{h}| \neq 0$ is satisfied when $\Phi < 1$ in the transition matrix. The state space model becomes unobservable in the equilibrium part $x_0(k)$, $x_1(k)$, and $x_2(k)$ for Φ approaching to 1. Theoretically, this means that a zero sampling time Δt is impossible within the particular model employed.

For Δt to infinity (i.e., Φ to zero), the computation of the inverse of the covariance matrix $P(k+1/k)$ by fixed-interval smoothing with the alternative state space model (Eq. 8.16) gives numerical problems, because two rows in $P(k+1/k)$ become almost identical resulting in a singular matrix. Here, one should switch to the state space model for a discrete titration (Eq. 8.3).

The measurement noise variance $R(k)$ is known reasonable well in advance. The diagonal element $Q_{44}(k-1)$ of the system noise covariance matrix $Q(k-1)$

is used as a tuning parameter to describe the monitored titration curve in a statistically correct way.

A new feature is the incorporation of first-order dynamics in the state space model. The exponential factor $\Phi = e^{-\Delta t/T_x}$ in the transition matrix $F(k, k-1)$ has to be known in advance and can be found by experimentation.

Note that here the transition matrix and measurement vector in the state space model contains only constants. Thus, the covariance matrices and associated gains in the Kalman filter and fixed-interval smoothing converge to a steady state.

In potentiometry, the relevant dynamics may involve the concentration and not the measured potential. If this is the case, we have to look for a model change. In practice, however, there may be no need for such a theoretical requirement.

8.2.2 Prediction

In this case, prediction is done by Eqs. (3.11) given in Section 3.3 with equidistant volume additions and thus uniformly spaced measurements.

8.2.3 Filtering

Filtering pertains to Eqs. (3.26) of the Kalman filter in Section 3.4.

8.2.4 Smoothing

Here, we are dealing with a fixed time delay and thus equidistant volume additions.

One employs Eqs. (3.28) in Section 3.6.1 for fixed-interval smoothing.

For online endpoint control (see Section 8.2.7), smoothing is of no use.

8.2.5 Evaluation

When the complete titration curve is processed, Eqs. (8.8) for the evaluated inflection point and/or setpoint are still valid. In the variance formula (8.9), one should use P_{44} instead of P_{33}.

The time constant T_x depends strongly on the chemical reaction, mixing of the solution, and the electrode response involved in the continuous titration system. In practice, the time constant is not known precisely and it may change slightly from titration to titration. When the overall time constant is not known correctly, the final evaluated results will be inaccurate. State estimation will produce a systematic error proportional to the difference between the implemented and real time constant.

In principle, the uncorrected inflection point(s) or setpoint(s) may be corrected afterward by the overall time constant. In this way, however, the titrimetric analysis becomes similar to a normal calibration procedure.

When first-order dynamics can be used for the measurements, online control to a preset endpoint in the titration curve is possible. When endpoint control in a continuous titration is applied, the volume counter of the burette is read as the evaluated result when the endpoint is reached.

8.2.6 Optimization

The transition matrix and measurement vector in the state space model contain only constants. Thus, the covariance matrices and associated gains of the Kalman filter and fixed-interval smoothing converge to a steady state. See further Section 8.1.6.

For state estimation, the optimal performance depends on the various terms in the state space model. Eq. (8.10) holds for the information yield I_{inf} of the evaluated inflection point. Here, one should use P_{44} instead of P_{33}. From Section 8.1.6, it follows that one has to minimize the measurement noise variance $R(k)$, system noise covariance matrix $Q(k-1)$ and/or volume step ΔV.

Now, consider Eq. (7.10) for the information yield $I(k)$. For an optimal solution, only the prediction part of the Kalman filter is of interest. The choice of the sampling time Δt affects both the exponential factor $\Phi = e^{-\Delta t/T_x}$ in the transition matrix $F(k, k-1)$ and the system noise covariance matrix $Q(k-1)$ by the relation $\Delta V = \rho \Delta t$. The total information yield $I_{tot}(k)$ pinpoints the system noise covariance matrix $Q(k-1)$ with a minimal volume step ΔV as the dominant factor. A minimal volume step ΔV is also dictated by the information yield I_{inf}. Therefore, with a fixed experimental time constant T_x the sampling time Δt and/ or the titrant rate ρ should be minimized for an optimal performance. But for Δt approaching to zero, i.e., Φ to 1, the state space model becomes unobservable somewhere.

Simultaneously, to minimize analysis time, the highest volume addition speed ρ possible should be selected. Thus, in practice, one has to find a compromise for the sampling time and volume addition speed to be applied within the given theoretical and experimental constraints.

8.2.7 Control

Online control can be applied toward an empirically established endpoint in the titration curve, in which the dynamics of the titration system are taken into account. The only way to manipulate the performance of the titration experimentally is by setting the burette either on or off in a given time transition. The Kalman filter is based on the given state space model (8.16) and the equation for endpoint control in an upward titration curve is described at index k by:

$$x_{max} = \hat{x}_0(k - 1/k - 1) + 3\sqrt{P_{22}(k - 1/k - 1)} \qquad (8.17)$$

$$x_{max} \geq x_{end} \Rightarrow \delta = 0 \quad (\text{burette off})$$

$$x_{max} < x_{end} \Rightarrow \delta = 1 \quad (\text{burette on})$$

$$z(k) \geq x_{end} \Rightarrow \text{stop} \quad (\text{endpoint reached})$$

where $\hat{x}_0()$ is taken from the estimate $\hat{\mathbf{x}}(k - 1/k - 1)$ and $P_{22}()$ is the second diagonal element of the covariance matrix $P(k - 1/k - 1)$. The expected value of x_{max} based on the estimated titration curve and its associated variance is compared with a preset endpoint x_{end} to be chosen. For a downward titration curve, Eq. (8.17) can be adjusted accordingly. To account for the variable volume addition in time, the term δ is included in the transition matrix:

$$F(k, k - 1) = \begin{pmatrix} \Phi & 1 - \Phi & (1 - \Phi)\delta & 1/2(1 - \Phi)\delta \\ 0 & 1 & \delta & 1/2\delta \\ 0 & 0 & 1 & \delta \\ 0 & 0 & 0 & 1 \end{pmatrix} \qquad (8.18)$$

The observability criterion $|M| \neq 0$ is satisfied when $\delta = 1$ but when $\delta = 0$ the state space model is not observable in the states $x_1(k)$ and $x_2(k)$. The Kalman filter cannot separate the observable part from the unobservable part in the estimated state. Therefore, some additional modifications are needed. The system noise covariance matrix with $Q_{44}(k - 1)\delta$ and the Kalman gain vector elements $k_3(k)\delta$ and $k_4(k)\delta$ are implemented. Now, the Kalman filter keeps the estimated states $\hat{x}_1(k/k)$ and $\hat{x}_2(k/k)$ constant if $\delta = 0$ and does not affect their variances.

8.2.8 Adaptation

See Section 8.1.8 on adaptation in discrete titrations. When online endpoint control (see Section 8.2.7) is employed, adaptation is of no use.

8.3 Nonlinear modeling

Here, one can formulate a nonlinear state space model with the exponential factor $\Phi(k)$ incorporated in the extended state, i.e., the state vector, measurement vector, and transition matrix are:

$$\mathbf{x}^T(k) = (x_d(k) \quad x_0(k) \quad x_1(k) \quad x_2(k) \quad \Phi(k)) \qquad (8.19)$$

$$\mathbf{h}^T(k) = (1 \quad 0 \quad 0 \quad 0 \quad 0)$$

$$f(\mathbf{x}(k - 1), k - 1) = F^*(\mathbf{x}(k - 1), k, k - 1)\mathbf{x}(k - 1)$$

$$F^*(\mathbf{x}(k-1), k, k-1) = \begin{pmatrix} \Phi & 1-\Phi & 1-\Phi & 1/2(1-\Phi) & 0 \\ 0 & 1 & 1 & 1/2 & 0 \\ 0 & 0 & 1 & 1 & 0 \\ 0 & 0 & 0 & 1 & 0 \\ 0 & 0 & 0 & 0 & 1 \end{pmatrix}$$

$$F(\mathbf{x}(k-1), k, k-1) = \frac{df(\mathbf{x}(k-1), k-1)}{d\mathbf{x}(k-1)} = \begin{pmatrix} \Phi & 1-\Phi & 1-\Phi & 1/2(1-\Phi) & a \\ 0 & 1 & 1 & 1/2 & 0 \\ 0 & 0 & 1 & 1 & 0 \\ 0 & 0 & 0 & 1 & 0 \\ 0 & 0 & 0 & 0 & 1 \end{pmatrix}$$

where $a = x_d(k-1) - x_0(k-1) - x_1(k-1) - 1/2x_2(k-1)$ and $\Phi = \Phi(k-1)$. With these definitions, both the extended Kalman filter and iterated linear filter-smoother operating with a linear measurement equation can be applied directly.

8.4 Simulation

The simulation of a titration curve is not as straightforward as for all the other examples in this book and needs some further explanation. The simulation of a titration curve comprises two stages: computation of the signal from a chemical system and the introduction of noise. At equilibrium, a chemical system is described by the equilibrium relations and the conservation laws for mass and charge. The following equation can be derived for a multicomponent acid/base system titrated with a strong base:

$$\sum_{i=1}^{A} \{V_s C_a(i)/[1+K_a(i)H]\} - \sum_{i=1}^{B} \{V_s C_b(i)K_b(i)H/[1+K_b(i)H]\}$$

$$+ (V_s + V_t)(K_w/H - H) - V_t C_t = 0 \tag{8.20}$$

where V_s is the original volume of the sample, V_t the volume of the titrant added, C_t the concentration of the titrant, A and B the number of acids and bases with $C_a()$ and $C_b()$ as their concentrations; K_w, $K_a()$, and $K_b()$ are equilibrium constants, and H is the hydrogen ion concentration.

The titration is monitored by potentiometry, the relevant form of the equilibrium pH_e being:

$$pH_e = -\log[H] \tag{8.21}$$

After addition of a small amount of titrant, the hydrogen ion concentration H in the system is calculated by an iterative Newton-Raphson procedure with the function value $f(H)$ and its first derivative $f'(H)$ based on Eq. (8.20):

$$H_{i+1} = H_i - f(H_i)/f'(H_i) \qquad (8.22)$$

until $|f(H_i)/H_{i+1} - f(H_i)/H_i| \leq \delta$ or $i \geq i_{max}$, where δ is a small value to be chosen and i_{max} the maximal number of iterations allowed. The computed hydrogen ion concentration is fed to Eq. (8.21) to obtain the required pH value.

The dynamic titration curve $pH_d(k)$ at index k is given by:

$$pH_d(k) = \Phi pH_d(k-1) + (1-\Phi)pH_e(k) \qquad (8.23)$$

There are two major sources of noise, one from the measurements and the other from the imprecision of the volume additions. These are incorporated in the simulation by addition of Gaussian white noise to the exact values. In order to produce the entire titration curve, this procedure is repeated for each volume addition.

As example, the titration of phosphoric and citric acid with a solution of sodium hydroxide was chosen. The data listed in Table 8.1 were used for simulation of the titration system.

8.5 Examples

Consider an acid-base titration followed by potentiometry of a weak acid with concentration 0.001 M and a sample volume of 50 mL. The titrant concentration of the strong base NaOH is 0.026 M. For phosphoric acid, the equivalence points are then 1.923 and 3.846 mL. For citric acid, the equivalence point is 5.769 mL.

Example 8.1: For a discrete titration, the titrant is added stepwise with equidistant volume additions of $\Delta V = 0.1$ mL and the final equilibrium pH's are the

TABLE 8.1 The data used for simulation of the titration system

Sample volume V_s		50 mL	
Titrant volume V_t		0–10 mL	
Titrant concentration C_t		0.026 M	
Phosphoric acid with $A = 3$, $B = 0$		**Citric acid with $A = 3$, $B = 0$**	
$C_a(i)$	$log(K_a(i))$	$C_a(i)$	$log(K_a(i))$
0.001 M	2.12	0.001 M	3.14
0.001 M	7.21	0.001 M	4.77
0.001 M	12.67	0.001 M	6.39

Noise variance $\sigma^2_{pH} = R$ and $\sigma^2_{V_t} = 0$ $log(K_w) = -14$

Exponential factor $\Phi = 0.9$

Newton-Raphson criteria $\delta = 10^{-10}$ and $i_{max} = 30$

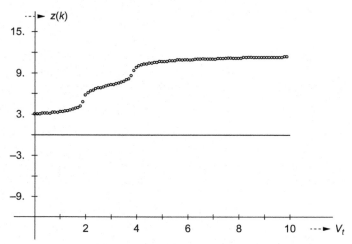

FIG. 8.1 Discrete titration system: the measurements with equidistant volume additions.

measurements. In Fig. 8.1 for phosphoric acid, the measurements with measurement noise and equidistant volume additions are shown. For state estimation, the state space model (8.3) is applied. The estimated titration curve and its derivatives by the Kalman filter are depicted in Fig. 8.2 and the same by fixed-interval smoothing in Fig. 8.3. For plotting purposes, the estimated state

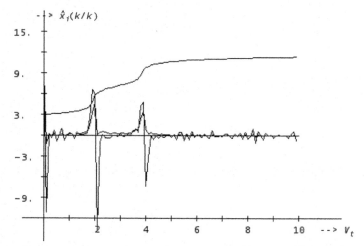

FIG. 8.2 The estimated state by the Kalman filter. $\hat{x}_0()$, $S\hat{x}_1()/\Delta V$, and $S^2\hat{x}_2()/\Delta V^2$ are plotted with a scaling factor $S = 0.4$.

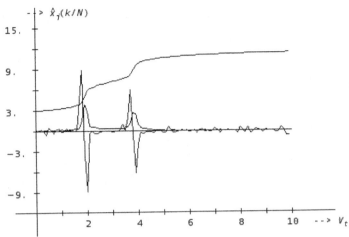

FIG. 8.3 The estimated state by fixed-interval smoothing (see Fig. 8.2).

is transformed according to the similarity matrix (8.4), i.e., $\hat{x}_0()$, $S\hat{x}_1()/\Delta V$, and $S^2\hat{x}_2()/\Delta V^2$ are plotted with the scaling factor $S = 0.4$.

The results for the evaluated setpoints and inflection points are given in Table 8.2. Here, also the implemented noise variances are shown. The Kalman filter has a systematic delay in the estimated state and so in the evaluated inflection points or setpoints, which is completely removed by subsequent fixed-interval smoothing.

For the setpoint volumes at pH 5 and pH 9, the results are less pronounced.

The corrected diagonal elements of the covariance matrix, i.e., $P_{11}()$, $S^2P_{22}()/\Delta V^2$, and $S^4P_{33}()/\Delta V^4$ according to the similarity matrix (8.4) are depicted in Fig. 8.4. It shows that the covariance matrix converges to a steady state.

Example 8.2: To provide equidistant measurements with variable volume additions, online control (8.11) can be integrated in the prediction and filtering equations of the Kalman filter. The same titration of phosphoric acid is considered. The volume unit for one-step forward in the discrete index variable is 0.01 mL. To do i steps in the sequence index, i volume units are added. A measurement stepsize $\Delta x_0 = 0.2$ pH units is used by the control algorithm with volume additions of minimal $\Delta V_{min} = 0.01$ mL and maximal $\Delta V_{max} = 0.25$ mL. Initially, two volume additions of $\Delta V_{init} = 0.1$ mL (i.e., three measurements) are done in order to get an observable estimate for a proper start of the online control algorithm. The equidistant measurements thus obtained with variable volume additions are shown in Fig. 8.5. Here, online control requires a total of 66 volume additions with corresponding measurements to do the job. There are a lot of measurements close to both the inflection points with observed minimal volume steps of 0.01 mL.

TABLE 8.2 Evaluated results in mL and settings for the described titration examples

Volumes in mL	pH 5	pH 9	V_{inf}	V_{inf}		
Phosphoric acid	1.923	3.846	1.923	3.846		
Example 8.1						
KF	1.913	3.839	2.080^a	3.990^a	$R=10^{-4}$	$Q_{33}=10^{-2}$
SM	1.908	3.835	1.922	3.842		
Example 8.2						
KF	1.914	3.836	1.960	3.878	$R=10^{-4}$	$Q_{33}=10^{-5}$
SM	1.912	3.836	1.925	3.845		
Citric acid	—	5.769	—	5.769		
Example 8.3						
KF	—	5.794	—	5.958	$R=10^{-5}$	$Q_{44}=10^{-2}$
SM	—	5.805	—	5.768	$\Phi=0.9$	
Example 8.4						
KF	—	5.8	—	—	$R=10^{-5}$	$Q_{44}=10^{-2}$
					$\Phi=0.9$	

50 mL of 0.001 M weak acid titrated with 0.026 M NaOH.
^{a}Volumes of the inflection points are corrected for a truncation error $1/2\Delta V$.

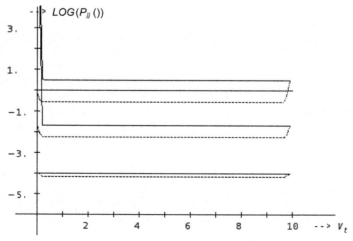

FIG. 8.4 The diagonal elements of the covariance matrix. $P_{11}()$, $S^2 P_{22}()/\Delta V^2$, and $S^4 P_{33}()/\Delta V^4$ are plotted with a scaling factor $S = 0.4$. (___) Kalman filter; (---) fixed-interval smoother.

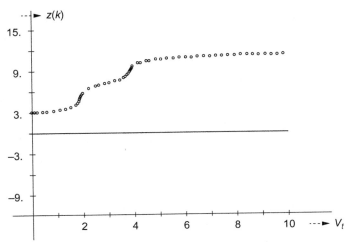

FIG. 8.5 Discrete titration system: the equidistant measurements with online volume control.

The estimated state by the Kalman filter and fixed-interval smoothing are depicted in Figs. 8.6 and 8.7. A scaling factor $S = 0.3$ and for ΔV the minimal volume step ΔV_{min} are used for plotting of the estimated state (see Example 8.1). The evaluated results can be found in Table 8.2. Again, the Kalman filter has a small systematic delay. Hereafter, fixed-interval smoothing gives precise and accurate results. The corrected diagonal elements of the covariance matrix are given in Fig. 8.8 (see Example 8.1), which demonstrates that both the

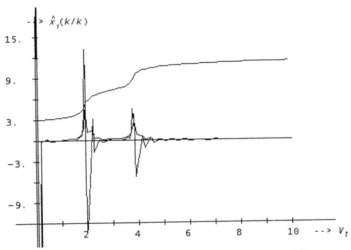

FIG. 8.6 The estimated state by the Kalman filter (see Fig. 8.2 with $S = 0.3$).

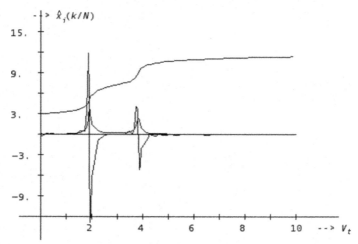

FIG. 8.7 The estimated state by fixed-interval smoothing (see Fig. 8.2 with $S = 0.3$).

estimation algorithms can be used with variable volume additions. Note that the covariance elements become smaller around the inflection points as should be.

In practice, only gross indications of the system and measurement noise variances $Q_{33}(k-1)$ and $R(k)$ have to be implemented in advance. After Kalman filtering with online volume control, fixed-interval smoothing provides almost always precise and accurate results.

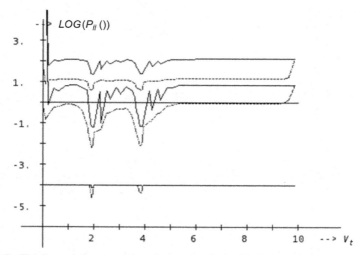

FIG. 8.8 The diagonal elements of the covariance matrix (see Fig. 8.4 with $S = 0.3$). (___) Kalman filter; (---) fixed-interval smoother.

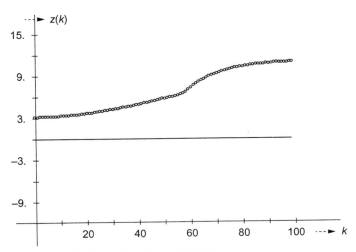

FIG. 8.9 Continuous titration system: the measurements.

Example 8.3: For a continuous titration, the volume step in one time unit is chosen as $\Delta V = \rho \Delta t = 0.1$ mL. We switch to the titration of the weak citric acid. The exponential factor $\Phi = 0.9$ is implemented in the state space model (8.16). This results in a time constant T_x of about 9.5 time units. In Fig. 8.9, the measurements generated with measurement noise are depicted. In Fig. 8.10, the estimated titration curve and its derivatives by the Kalman filter are shown and the same by fixed-interval smoothing in Fig. 8.11. For plotting the estimated state

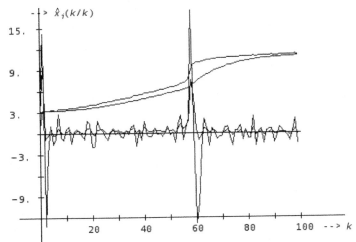

FIG. 8.10 The estimated state by the Kalman filter. $\hat{x}_d()$, $\hat{x}_0()$, $S\hat{x}_1()/\Delta V$, and $S^2\hat{x}_2()/\Delta V^2$ are plotted with $S = 0.5$.

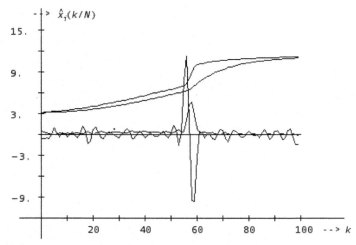

FIG. 8.11 The estimated state by fixed-interval smoothing (see Fig. 8.10).

$\hat{x}_d()$, $\hat{x}_0()$, $S\hat{x}_1()/\Delta V$, and $S^2\hat{x}_2()/\Delta V^2$ with a scaling factor $S = 0.5$ is used. The corrected diagonal elements $P_{11}()$, $P_{22}()$, $S^2 P_{33}()/\Delta V^2$, and $S^4 P_{44}()/\Delta V^4$ of the covariance matrix are depicted in Fig. 8.12. It shows that the covariance matrix converges to a steady state. Evaluated results can be found in Table 8.2 and the conclusions are similar as before. In practice, the volume step in one time unit can be chosen to be smaller. The constraint is governed by the observability

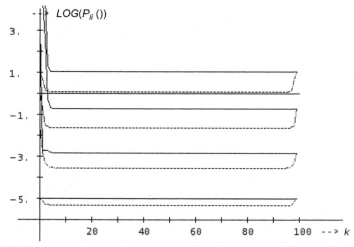

FIG. 8.12 The diagonal elements of the covariance matrix. $P_{11}()$, $P_{22}()$, $S^2 P_{33}()/\Delta V^2$, and $S^4 P_{44}()/\Delta V^4$ are plotted with a scaling factor $S = 0.5$. (___) Kalman filter; (---) fixed-interval smoother.

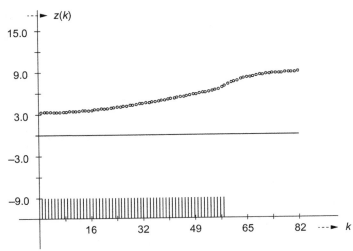

FIG. 8.13 Continuous titration system: the measurements with online endpoint control; *small lines* below denote burette on/off.

criterion, because for $\Delta t \rightarrow 0$ and thus $\Phi \rightarrow 1$ the system becomes unobservable somewhere.

Example 8.4: When a first-order response is used for the measurements, online control to an empirically established endpoint in the titration curve can be accomplished by applying Eq. (8.17). The same titration as in Example 8.3 was tested with burette control to a preset endpoint pH 9. The volume step in one time unit is chosen as $\Delta V = \rho \Delta t = 0.1$ mL and the exponential factor $\Phi = 0.9$.

Fig. 8.13 shows the performance of the measured pH's with the burette settings on or off. After the titration has started, the burette is kept on because the measurements are still far from the established endpoint. The burette is closed when the endpoint is finally reached. Fig. 8.14 shows the estimated state by the Kalman filter. For plotting of the estimated state, a scaling factor $S = 0.25$ is employed (see Example 8.3). When the burette is off, the estimated derivatives remain constant. The finally reached endpoint is 5.8 mL, which is in this case close enough to the theoretical equivalence volume.

In practice, the volume step in one time unit can be chosen to be smaller keeping in mind that the system becomes unobservable for $\Delta t \rightarrow 0$ and thus $\Phi \rightarrow 1$.

Only gross indications of the system noise variance $Q_{44}(k-1)$, measurement noise variance $R(k)$, and exponential factor Φ have to be known for online endpoint control. Especially, in case of a slow response and/or a steep inflection, there are advantages in both accuracy and analysis time when the proposed control algorithm is applied.

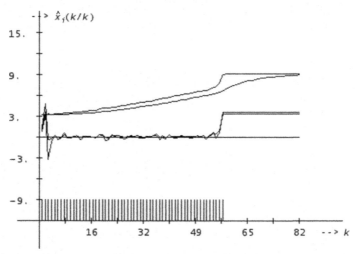

FIG. 8.14 The estimated state by the Kalman filter (see Fig. 8.10 with $S = 0.25$); *small lines* below denote burette on/off.

Example 8.5: An attempt is made to estimate in a nonlinear fashion simultaneously the exponential factor with the titration curve and its derivatives. The same titration as in Example 8.3 with the measurements from Fig. 8.9 is considered. The implemented nonlinear state space model is given in Section 8.3. The extended Kalman filter ($\ell=1$) and the iterated linear filter-smoother ($\ell=3$), both operating with a linear measurement equation, are applied. It is worth to mention that for this example state estimation was not very successful. Starting from different values of $\hat{\Phi}(0/0)$ and tuning parameters, the exponential factor $\hat{\Phi}(k/k)$ in the estimated state does not converge always properly to its right value and it looks like the system is unobservable. Therefore, no results are presented for this example.

Example 8.6: Furthermore, an attempt to estimate by multiple modeling the exponential factor in addition to the titration curve and its derivatives was also unsuccessful (see Section 9.1). The same titration as in Example 8.3 with the measurements from Fig. 8.9 is considered. Here, the unknown parameter vector $\mathbf{a}_i$ equals the exponential factor Φ_i in the transition matrix $F(k,k-1)$ of the applied state space model (8.16), i.e., $\mathbf{a}_i = \Phi_i$, $i = 1,2,\ldots,\ell$, where ℓ denotes the number of models employed. No results are given for this example.

Chapter 9

Miscellaneous

Chapter outline

9.1 Multiple modeling

The purpose of multiple modeling is to compute the conditional probability density:

$$p\{\mathbf{x}, \mathbf{a} \mid Z(k)\} = p\{\mathbf{x} \mid \mathbf{a}, Z(k)\} p\{\mathbf{a} \mid Z(k)\} \qquad (9.1)$$

where $\mathbf{a}$ denotes a vector of uncertain parameters somewhere in the state space model.

For a state space model, the first density on the right side of this expression is assumed to be Gaussian, with the mean $\hat{\mathbf{x}}(k/k)$ and covariance matrix $P(k/k)$ as computed by a Kalman filter, for each given value of the parameter vector $\mathbf{a}$. To enhance feasibility, let the parameter vector $\mathbf{a}_i$ assume to have only a finite number of values $i = 1, 2, \dots, \ell$.

One can now seek an online algorithm to produce the true mean and covariance matrix of the state with identification of the correct parameter vector. Now, redefine the hypothesis probability as the identity $p_i(k) \equiv p\{\mathbf{a}_i \mid Z(k)\}$. This gives the following equations for the mean state estimate and its covariance matrix:

$$\hat{\mathbf{x}}(k/k) = E\{\mathbf{x}(k) \mid Z(k)\} = \sum_{i=1}^{\ell} p_i(k) \hat{\mathbf{x}}_i(k/k) \qquad (9.2)$$

$$P(k/k) = E\left\{ [\mathbf{x}(k) - \hat{\mathbf{x}}(k/k)][\mathbf{x}(k) - \hat{\mathbf{x}}(k/k)]^{\mathrm{T}} \mid Z(k) \right\}$$

$$= \sum_{i=1}^{\ell} p_i(k) \left\{ P_i(k/k) + [\hat{\mathbf{x}}_i(k/k) - \hat{\mathbf{x}}(k/k)][\hat{\mathbf{x}}_i(k/k) - \hat{\mathbf{x}}(k/k)]^{\mathrm{T}} \right\} \qquad (9.3)$$

$$p_i(k) = \frac{p\{z(k) \mid \mathbf{a}_i, Z(k-1)\} p_i(k-1)}{\sum\limits_{j=1}^{\ell} p\{z(k) \mid \mathbf{a}_j, Z(k-1)\} p_j(k-1)} \quad \text{for } i = 1, 2, \dots, \ell \qquad (9.4)$$

State Estimation in Chemometrics. https://doi.org/10.1016/B978-0-08-102603-8.00009-2

where $\hat{\mathbf{x}}_i(k/k)$ and $P_i(k/k)$ are the state estimate and covariance matrix produced by a Kalman filter based on the assumption that the parameter vector equals $\mathbf{a}_i$. The mean state estimate $\hat{\mathbf{x}}(k/k)$ is the weighted average of the state estimates generated by each of the ℓ separate Kalman filters using the probabilities $p_i(k)$. The mean covariance matrix $P(k/k)$ is similarly based upon the batch of Kalman filters. $P(k/k)$ cannot be precomputed since it involves $p_i(k)$, $\hat{\mathbf{x}}_i(k/k)$, and $\hat{\mathbf{x}}(k/k)$, all of which require knowledge of the measurement history.

When the measurement $z(k)$ becomes available, the innovations $v_i(k)$ are generated by the ℓ filters and passed on for processing the hypothesis probability computation. Specifically, each $p\{z(k)|\mathbf{a}_i, Z(k-1)\}$, $i = 1, 2, \ldots, \ell$ can be computed:

$$p\{z(k)|\mathbf{a}_i, Z(k-1)\} = \frac{1}{\{2\pi|A_i(k)|\}^{1/2}} e^{-1/2 v_i^{\mathrm{T}}(k) A_i^{-1}(k) v_i(k)} \qquad (9.5)$$

$$\text{with}: \quad v_i(k) = z(k) - \mathbf{h}_i^{\mathrm{T}}(k)\hat{\mathbf{x}}_i(k/k-1)$$

$$A_i(k) = \mathbf{h}_i^{\mathrm{T}}(k)P_i(k/k-1)\mathbf{h}_i(k) + R_i(k)$$

Prior to processing any measurements, one must initialize the probabilities $p_i(0)$. If all models are equally likely, then $p_i(0) = 1/\ell$ for $i = 1, 2, \ldots, \ell$. The ℓ computations, along with the memorized previous probabilities $p_i(k-1)$, allow the computation of the current probabilities $p_i(k)$. One could expect that the innovations of the Kalman filter based upon the "*correct*" model will be consistently smaller than the ones from the mismatched filters.

If it is desired to produce an estimate of the parameter vector $\hat{\mathbf{a}}(k/k)$:

$$\hat{\mathbf{a}}(k/k) = E\{\mathbf{a}|Z(k)\} = \sum_{i=1}^{\ell} p_i(k)\mathbf{a}_i \qquad (9.6)$$

$$E\left\{[\mathbf{a} - \hat{\mathbf{a}}(k/k)][\mathbf{a} - \hat{\mathbf{a}}(k/k)]^{\mathrm{T}}|Z(k)\right\} = \sum_{i=1}^{\ell} p_i(k)[\mathbf{a}_i - \hat{\mathbf{a}}(k/k)][\mathbf{a}_i - \hat{\mathbf{a}}(k/k)]^{\mathrm{T}}$$

We are now in position to reflect the whole multiple modeling system in perspective. As measurements are accumulated, each individual Kalman filter determines its weighted squared innovation, uses this in the computation of the probability density function value, and thereafter the new probabilities are determined. Finally, the mean state estimate and its covariance matrix are computed. As the index k increases, for the correct model, innovations work out to be smaller than the others and this gives the most weight in the blending of the individual estimates. The innovations are summed in recursive form and finally in the limit a probability of unity (one) is found for the correct model.

Optimization can be done similar to the optimal design computation in the multicomponent system. For the expected measurement and its variance hold:

$$E\{z(k)|Z(k-1)\} = \hat{z}(k) = \sum_{i=1}^{\ell} p_i(k-1)\hat{z}_i(k) \quad \text{with} \quad \hat{z}_i(k) = \mathbf{h}_i^{\mathrm{T}}(k)\hat{\mathbf{x}}_i(k/k-1)$$

(9.7)

$$E\left\{(z(k)-\hat{z}(k))(z(k)-\hat{z}(k))^{\mathrm{T}}|Z(k-1)\right\}$$
$$= \sum_{i=1}^{\ell} p_i(k-1)\left\{(\hat{z}_i(k)-\hat{z}(k))^2 + \mathbf{h}_i^{\mathrm{T}}(k)P_i(k/k-1)\mathbf{h}_i(k) + R_i(k)\right\}$$

If the system equation in the state space model is index invariant, the following formula holds for optimization:

Maximize $\displaystyle\sum_{i=1}^{\ell} p_i(k-1)\left\{(\hat{z}_i(k)-\hat{z}(k))^2/R_i(k) + \mathbf{h}_i^{\mathrm{T}}(k)P_i(k/k-1)\mathbf{h}_i(k)/R_i(k)\right\}$

(9.8)

Here, optimization can only be applied online, since the probabilities $p_i(k-1)$ and the predicted states $\hat{\mathbf{x}}_i(k/k-1)$ require knowledge of the measurement history.

During the progress of multiple modeling, the algorithm switches automatically from the modeling to the estimation mode.

The measuring process can be terminated when a preset imprecision is reached.

To check the correctness and validity of the right model, the online means and chi-square values of the standardized innovations for each model should be computed for verification. If the means and/or chi-square values exaggerate for the best model determined from the measurements, it is not included in the model database.

9.2 Principal components

Assume the samples $Z(k) = \{\mathbf{z}(1), \mathbf{z}(2), \ldots, \mathbf{z}(k)\}$ are available, where $\mathbf{z}(i)$ is an m-column vector that contains measurements, which are linear and additive composed. The correlation matrix of the first k samples is:

$$R(k) = \frac{1}{k}Z(k)Z^{\mathrm{T}}(k) = \frac{1}{k}\sum_{i=1}^{k}\mathbf{z}(i)\mathbf{z}^{\mathrm{T}}(i)$$

(9.9)

where $Z(k)$ is an $m*k$ matrix and $R(k)$ is a symmetric positive definite $m*m$ matrix.

By expressing this in a recursive form, the following online scheme for modeling is obtained for computation of the n largest eigenvectors of $R(k)$:

$$R(k) = (1-1/k)R(k-1) + 1/k\{\mathbf{z}(k)\mathbf{z}^{\mathrm{T}}(k)\}$$

(9.10)

$$\mathbf{v}_i(k) = R(k)\mathbf{u}_i(k-1) \quad \text{for } i = 1, 2, \ldots, n \tag{9.11}$$

$$\mathbf{w}_i(k) = \left[I - \sum_{j=1}^{i-1} \mathbf{u}_j(k)\mathbf{u}_j^{\mathrm{T}}(k)\right]\mathbf{v}_i(k) \quad (\mathbf{w}_1(k) = \mathbf{v}_1(k)) \tag{9.12}$$

$$\mathbf{u}_i(k) = \mathbf{w}_i(k)/\|\mathbf{w}_i(k)\| \tag{9.13}$$

$\mathbf{u}_i(k)$, $\mathbf{v}_i(k)$, and $\mathbf{w}_i(k)$ are m-column vectors and "$\|\ \|$" is the Euclidian norm.

Effectively (9.11) is the power step, which multiplies the eigenvector estimate $\mathbf{u}_i(k-1)$ by the matrix $R(k)$. Steps (9.12) and (9.13) simply orthonormalize the multiplied vectors $\mathbf{v}_i(k)$ using the well-known classical Gram-Schmidt procedure. Alternatively, the modified Gram-Schmidt computation can be applied.

If the n largest eigenvalues of $R(k)$ are distinct, then the estimate $\mathbf{u}_i(k)$ converges to the ith largest eigenvector as k grows large enough.

The estimated eigenvectors $\mathbf{u}_i(k)$; $i = 1, 2, \ldots, n$ are collected in the $m*n$ eigenvector matrix $U(k)$.

The scalar eigenvalues $\lambda_i(k)$ can be found easily either by:

$$\lambda_i(k) = \|\mathbf{w}_i(k)\| \quad \text{or}: \quad \lambda_i(k) = \mathbf{u}_i^{\mathrm{T}}(k)R(k)\mathbf{u}_i(k) \tag{9.14}$$

An even faster online algorithm for principal component analysis is obtained by assuming that:

$$R(k-1)\mathbf{u}_i(k-1) \approx \lambda_i(k-1)\mathbf{u}_i(k-1) \tag{9.15}$$

This means that the estimate $\mathbf{u}_i(k-1)$ is roughly an eigenvector of the matrix $R(k-1)$ with $\lambda_i(k-1)$ being its eigenvalue. Using (9.15), Eqs. 9.10 and 9.11 combined take the form:

$$\mathbf{v}_i(k) = \left[I + \frac{1}{(k-1)\lambda_i(k-1)}\mathbf{z}(k)\mathbf{z}^{\mathrm{T}}(k)\right]\mathbf{u}_i(k-1) \tag{9.16}$$

The scalar coefficient $(k-1)\lambda_i(k-1)/k$ in front of the square brackets can be omitted because of the normalization contained in (9.12) and (9.13).

The eigenvalue needed in (9.16) can be estimated recursively by:

$$\lambda_i(k) = (1 - 1/k)\lambda_i(k-1) + 1/k\{\mathbf{z}^{\mathrm{T}}(k)\mathbf{u}_i(k-1)\}^2 \tag{9.17}$$

Note that in Eqs. (9.16) and (9.17) only vector products are needed because the product $\mathbf{z}^{\mathrm{T}}(k)\mathbf{u}_i(k-1)$ is computed first. If the dimensionality of the measurements is large and only a few eigenvectors are needed, considerable savings in computation time can be achieved by applying formulas (9.16) and (9.17) instead of (9.10), (9.11), and (9.14).

For online evaluation, the number of significant principal components follows from the estimated eigenvalues by the statistical formula:

$$F(1, m-i) = \frac{\lambda_i(k)(m-i)}{\sum\limits_{j=i+1}^{m} \lambda_j(k)} = \frac{\lambda_i(k)(m-i)}{tr\{R(k)\} - \sum\limits_{j=1}^{i} \lambda_j(k)} \quad i = n, n-1, \ldots, 1 \quad (9.18)$$

where $F(1, m-i)$ is the Fisher F-value to be tested one-sided against a critical statistical table value for a given confidence level α and $(1, m-i)$ degrees of freedom. Here, i is indexing in the reverse way until the computed F-value differs significantly. For the trace of $R(k)$, i.e., $tr\{R(k)\}$ holds in recursive form simply:

$$tr\{R(k)\} = (1 - 1/k)tr\{R(k-1)\} + 1/k\{\mathbf{z}^T(k)\mathbf{z}(k)\} \quad (9.19)$$

The principal components modeled online after k measurements by Eqs. (9.11)–(9.14) can be replaced and improved by a local iteration for $j = 1, 2, \ldots, \ell$:

$$V^j(k) = R(k)U^{j-1}(k) \quad (9.20)$$

$$V^j(k) = U^j(k)S^j(k) \quad (9.21)$$

where (9.20) denotes the power step and (9.21) the Gram-Schmidt orthogonalization either in classical or in modified form. $S^j(k)$ is an upper triangular $n*n$ matrix with on the diagonal the positive eigenvalues $\lambda_i(k)$ computed such that the $m*n$ matrix $U^j(k)$ has orthonormal columns $\mathbf{u}_i^j(k)$, $i = 1, 2, \ldots, n$.

Until now, the orthonormal decomposition describes the eigenvectors $\mathbf{u}_i(k)$; $i = 1, 2, \ldots, n$ of $1/kZ(k)Z^T(k)$ (row mode) contained in the matrix $U(k)$. The k-column vectors $\mathbf{q}_i(k)$ in the $k*n$ matrix $Q(k)$, which are the eigenvectors of $1/mZ^T(k)Z(k)$ (column mode), are found by the relationship for $i = 1, 2, \ldots, n$:

$$\mathbf{q}_i(k) = \frac{1}{\sqrt{k\lambda_i(k)}} Z^T(k)\mathbf{u}_i(k) \quad (9.22)$$

In this way, one can apply either target transformation factor analysis directly or iteratively on the eigenvectors $U(k)$ or $Q(k)$.

9.3 Standard addition

The quality of calibration by most analytical methods fixes the quality of the final results. In most cases, there is no absolute quantitative relationship between analyte concentration and response, so that the measurement obtained for an unknown sample has to be compared with the measurements of a number of calibration samples. The problems associated with calibration are well known: the selectivity of many analytical samples is poor, and many constituents present in the sample may affect the measurements, i.e., the interference effects. A second serious problem associated with calibration is the varying sensitivity of the sensor with a changing matrix in the sample, i.e., the matrix effects. Of course, calibration methods, which handle these problems adequately, have been available for many years. Interference effects may be compensated by applying multicomponent analysis. Matrix effects can be avoided by using the standard addition method.

Until the late 1970s, systems where interference and matrix effects occur simultaneously could not be calibrated in a direct way. The problem has been solved by combining the multicomponent analysis and standard addition method into a widely applicable method, the generalized standard addition method (GSAM). The benefit of GSAM is not only the assessment of calibration in a more general way, but it provides also a relationship between the imprecision of the analytical result and the measurement error, in relation to the experimental design.

With GSAM multiple components can be analyzed at once. The strategy to follow covers the use of p multiple sensors and the processing of n volume additions for each of the r components. In order to get a set of solvable equations, it is necessary that the number of sensors is greater or equal to the number of components to analyze, i.e., $p \geq r$. We have to approach the problem mathematically as if each component delivers a contribution to the measurement of the ith sensor:

$$r_i = \sum_{j=1}^{r} c_j k_{ji} + e_i \qquad (9.23)$$

where r_i is the measurement, c_j the concentration involved, e_i the added noise, and k_{ji} is the sensitivity of analyte j at sensor i.

If the measurements of all sensors are combined, this gives the following equation:

$$\mathbf{r}^{\mathrm{T}} = \mathbf{c}^{\mathrm{T}} K + \mathbf{e}^{\mathrm{T}} \qquad (9.24)$$

where $\mathbf{r}$ is an p-column vector containing the measurements, $\mathbf{c}$ an r-column vector with the concentrations involved, $\mathbf{e}$ an p-column vector with the added noise, and K is an $r*p$ matrix with the selected sensitivities.

In the first instance, there are n volume additions done for all the r components with $n \geq r$ in order to estimate the sensitivity matrix K by means of classical least squares. Therefore, added amounts or number of moles $\Delta \mathbf{n}_k$ are introduced with k as the kth volume addition. For the measurements, a volume correction with use of difference vectors has to be applied by substracting the corrected responses at $k = 0$. For the corrected measurements $\Delta \mathbf{q}_k$ follows then:

$$\Delta \mathbf{q}_k^{\mathrm{T}} = \Delta \mathbf{n}_k^{\mathrm{T}} K + \mathbf{e}_k^{\mathrm{T}} \qquad (9.25)$$

In fact, the noise vector $\mathbf{e}_k$ in Eq. (9.25) is not equal to the $\mathbf{e}$ in (9.24), and it is easier to treat them in both equations similarly. If the measurements for n volume additions are collected, we get the following matrix equation:

$$\Delta Q = \Delta N K + E \qquad (9.26)$$

Here ΔQ and E are $n*p$ matrices, ΔN is an $n*r$ matrix, and K is an $r*p$ matrix. With n as the number of volume additions ΔQ, ΔN and E are defined by:

$$\Delta Q = \begin{pmatrix} \Delta \mathbf{q}_1^{\mathrm{T}} \\ \Delta \mathbf{q}_2^{\mathrm{T}} \\ \cdot \\ \cdot \\ \cdot \\ \Delta \mathbf{q}_n^{\mathrm{T}} \end{pmatrix} \qquad \Delta N = \begin{pmatrix} \Delta \mathbf{n}_1^{\mathrm{T}} \\ \Delta \mathbf{n}_2^{\mathrm{T}} \\ \cdot \\ \cdot \\ \cdot \\ \Delta \mathbf{n}_n^{\mathrm{T}} \end{pmatrix} \qquad E = \begin{pmatrix} \mathbf{e}_1^{\mathrm{T}} \\ \mathbf{e}_2^{\mathrm{T}} \\ \cdot \\ \cdot \\ \cdot \\ \mathbf{e}_n^{\mathrm{T}} \end{pmatrix} \qquad (9.27)$$

If Eq. (9.26) consists of column vectors instead of matrix equations with exception of K, then the classical least squares approach can be applied by minimizing the Euclidian norm $\|E\|$:

$$\|E\| = \sum_i e_i^2 = E^{\mathrm{T}} E \qquad (9.28)$$

In analogy, a similar approach can be applied, now with $E^{\mathrm{T}}E$ as a matrix from Eq. (9.26).

$$E = \Delta Q - \Delta N K \qquad (9.29)$$

Thus: $E^{\mathrm{T}} E = \Delta Q^{\mathrm{T}} \Delta Q - \Delta Q^{\mathrm{T}} \Delta N K - K^{\mathrm{T}} \Delta N^{\mathrm{T}} \Delta Q + K^{\mathrm{T}} \Delta N^{\mathrm{T}} \Delta N K \qquad (9.30)$

Partial differentiation with respect to K and setting the result equal to zero gives:

$$-2\Delta N^{\mathrm{T}} \Delta Q + 2\Delta N^{\mathrm{T}} \Delta N K = 0 \qquad (9.31)$$

From which follows immediately the solution for classical least squares:

$$K = \left(\Delta N^{\mathrm{T}} \Delta N \right)^{-1} \Delta N^{\mathrm{T}} \Delta Q \qquad (9.32)$$

The first step in the computation is now completed: there is an estimate of the sensitivity matrix K. With an estimated K, the r-column vector $\mathbf{n}_0$ can be evaluated by using the p-column vector of the volume corrected measurements $\mathbf{q}_0$ by solving:

$$\mathbf{q}_0 = K^{\mathrm{T}} \mathbf{n}_0 \qquad (9.33)$$

The vector $\mathbf{n}_0$ is given by:

$$\mathbf{n}_0 = \left(K K^{\mathrm{T}} \right)^{-1} K \mathbf{q}_0 \qquad (9.34)$$

For clarity, the application of the volume correction on the measurements has no influence on the estimated K in Eq. (9.34). Therefore, $\mathbf{r}_0$ can be applied instead of $\mathbf{q}_0$ from which $\mathbf{c}_0$ can be calculated immediately.

Optimization of the experimental design used in Eq. (9.34), i.e., the optimal selection of wavelenghts in spectroscopy can be done by measuring the spectra of the pure components assuming that these spectra are only slightly different in the sample matrix. Thereafter, one uses the recursive approach for the computation of the optimal design as described in Section 6.1.5.2.

Now follows the description of a recursive approach for the online estimation of the sensitivities available in the system under investigation. The problem of calibration at one sensor will discussed first. For a sample with r analytes, the initial measurement before volume addition at sensor i is:

$$r_{0i} = c_{01}k_{1i} + c_{02}k_{2i} + \cdots + c_{0r}k_{ri} + e_i \tag{9.35}$$

Because the number of moles is additive, instead of concentrations, volume corrected measurements and amounts are introduced: $q_{0i} = r_{0i}V_0$ and $n_{0i} = c_{0i}V_0$, where V_0 is the initial volume of the sample. Eq. (9.35) then becomes:

$$q_{0i} = n_{01}k_{1i} + n_{02}k_{2i} + \cdots + n_{0r}k_{ri} + e_i \tag{9.36}$$

$$\text{or in vector notation}: \quad q_{0i} = \mathbf{n}_0^T \mathbf{k}_i + e_i \tag{9.37}$$

The vector of the sensitivities $\mathbf{k}_i$ will be estimated in a recursive procedure. If all the analytes are added simultaneously with a total volume addition Δv_k and the added number of moles of analyte i is Δn_{ki}, then the change of the measurement at sensor i after volume addition k is:

$$\Delta q_{ki} = r_{ki}(V_0 + \Delta v_k) - q_{0i} = \begin{pmatrix} \Delta n_{k1} & \Delta n_{k2} & \ldots & \Delta n_{kr} \end{pmatrix} \begin{pmatrix} k_{1i} \\ k_{2i} \\ \vdots \\ \vdots \\ k_{ri} \end{pmatrix} = \Delta \mathbf{n}_k^T \mathbf{k}_i + e_i \tag{9.38}$$

With this equation, recursive least squares or the Kalman filter can be applied directly for sensor i. Note that the system equation is index invariant, i.e., the transition matrix $F(k, k-1) = I$ and the system noise $\mathbf{w}(k-1) = 0$.

The recursive estimation procedure runs parallel separately for each of the sensors $i = 1, 2, \ldots, p$ and has to be processed again for every volume addition $k = 1, 2, \ldots, n$.

GSAM constitutes a first step toward a self-calibrating system, which will become of full value when the measurement is placed with the progress of the experimental design. This means that each step in the calibration stage is decided on the basis of all measurements collected previously. In its classical form as described before, GSAM does not provide this feature. All volume additions precede the calculation of the sensitivities, the analyte concentrations, and the propagated error in the final results. This implies that there is no check that the underlying model remains valid during the volume additions. No indication is obtained on the necessity of further additions, taken into account the desired imprecision of the analytical result. For this purpose, an algorithm is needed, which returns improved least squares estimates of the sensitivities, every time a new measurement becomes available. Such a recursive algorithm opens the way to achieve a feedback between response and experimental design, which is essential in the development of a self-calibration system.

The recursive version of GSAM adds new powerful properties to the method:

(1)   Online control of the validity and correctness of the linear model during calibration. The innovation can be used in a warning system for a faulty mode and can be achieved by a sequential test. One could expect that the online means and chi-square values of the standardized innovations would do the job.

However, because of the transformation of the measurements by volume correction and using difference values, the measurement noise variance cannot be used. One has to deal with an adjusted noise variance for the corrected measurements. Because of the use of difference measurements, the standardized innovations are no longer Gaussian white noise anymore.

A test on the averaged innovation is very similar to detecting drift in a random test series by examining the cumulative sum of the innovations. Further widely used methods for deciding whether or not a trend is significant are the CUSUM-test, V-mask and Trigg's monitoring technique.

(2)   Simultaneously evaluation of the value and imprecision of the sensitivity vector during the volume additions. As soon as the estimate of the sensitivities converges to a steady state, no further improvement obtained can be expected and the volume additions at that sensor can reasonable be stopped. A criterion for stopping the volume additions may be found by monitoring the diagonal elements of the covariance matrix, which indicates the imprecision of the estimated sensitivities after each volume addition. When the imprecision has reached the desired limits as a proper analytical goal, no further volume additions are necessary.

(3)   Access to an online interactive calibration, terminating the volume additions, when either the imprecision is within desired limits, or a deficiency in the model has been detected. This is further quite similar as described above.

9.4   Examples

Example 9.1: The first example of multiple modeling is directly taken from multicomponent analysis in spectroscopy. Consider a one-component system with the individual molar spectra given in Fig. 9.1 and $\ell = 10$ candidate components. Each one-component model is described by the state space model with a system equation, which is index invariant and a given measurement equation for $i = 1,2,\ldots,\ell$:

$$z(k) = (\, h_i(k) \quad 1\,) \begin{pmatrix} c_i(k) \\ b_i(k) \end{pmatrix} + v(k) \tag{9.39}$$

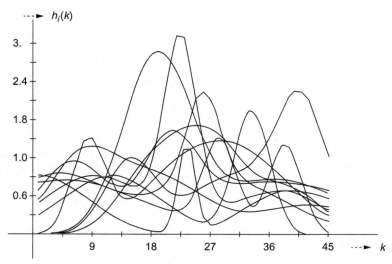

FIG. 9.1 Multiple modeling: the molar spectra of the pure components.

where $h_i(k)$ are the molar absorptivities of the pure components at a given wavelength k, $c_i(k)$ is the concentration, and $b_i(k)$ the baseline involved to be estimated for the ith model.

Here, $\mathbf{a}_i$ denotes simply the ith candidate model with $i = 1, 2, \ldots, \ell$.

In Fig. 9.2, the measurements with a noise variance $R(k) = 10^{-5}$ employed are depicted. Performing multiple modeling processes these measurements. The

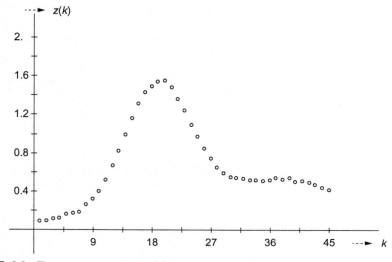

FIG. 9.2 The measurements involved for one component.

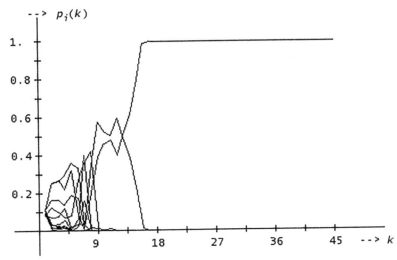

FIG. 9.3   The probabilities for each model: forward scan.

results are shown in Fig. 9.3 where the probabilities are plotted for forward scanning through the measurements. In Fig. 9.4, the mean state estimate is shown. For the final mean state estimate, the values of the concentration and baseline should be 0.5 and 0.1, respectively. Both figures demonstrate that after about 20 measurements the system has converged to the one right model with its corresponding state estimate.

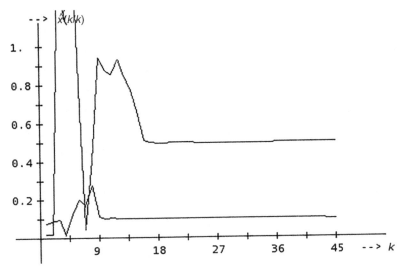

FIG. 9.4   The mean state estimate: forward scan.

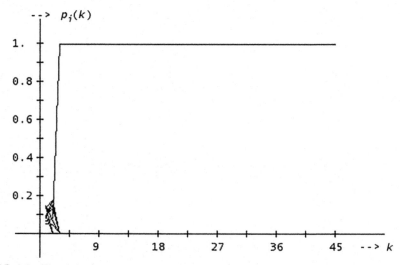

FIG. 9.5   The probabilities for each model: online optimal selection.

In Figs. 9.5 and 9.6, similar results are shown for the online optimal selection of wavelengths processing the measurements. Here, the right model and its estimated state are already found within five measurements.

Examination of the online means and chi-square values based on the standardized innovations indicate in both cases that the determined best model is indeed a correct and valid one.

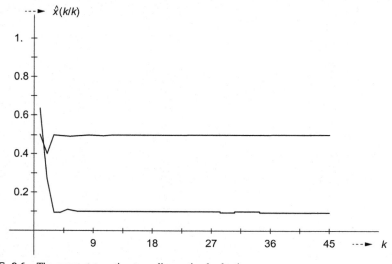

FIG. 9.6   The mean state estimate: online optimal selection.

The state space model (9.39) can be extended to cover more components and/or to include stochastic drift.

Example 9.2: The second example of multiple modeling uses the same individual molar spectra of the pure components as in Fig. 9.1 with $\ell = 10$. Here, a three-component system is considered. Now, $\mathbf{a}_i$ denotes the ith candidate model with $i = 1,2,...,p$, where p is defined by the total number of possible combinations $p = \frac{\ell!}{(\ell-n)!n!}$. The number of components n in the state space model equals $n = 3$. Thus, for 3 components in the state space model and 10 individual molar spectra, there are $p = 120$ possible combinations.

The three-component system is given by the state space model with a system equation that is index invariant and the measurement equation becomes for $i = 1,2,...,p$:

$$z(k) = \left(h_{i1}(k) \quad h_{i2}(k) \quad h_{i3}(k) \quad 1 \right) \begin{pmatrix} c_{1i}(k) \\ c_{2i}(k) \\ c_{3i}(k) \\ b_i(k) \end{pmatrix} + v(k) \qquad (9.40)$$

where $h_{i1}(k)$, $h_{i2}(k)$, and $h_{i3}(k)$ are the molar absorptivities of the pure components at a given wavelength k; $c_{1i}(k)$, $c_{2i}(k)$, and $c_{3i}(k)$ are the concentrations and $b_i(k)$ the baseline involved to be estimated for the ith model $i = 1,2,...,p$.

In Fig. 9.7, the measurements with a noise variance $R(k) = 10^{-5}$ employed are depicted. Performing multiple modeling processes these measurements. The results are shown in Fig. 9.8 where the probabilities are plotted for forward scanning through the measurements. In Fig. 9.9, the mean state estimate is

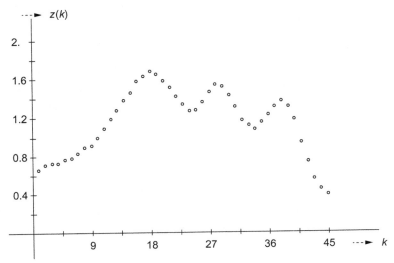

FIG. 9.7 Multiple modeling: the measurements involved for three components.

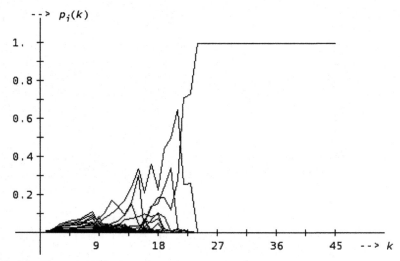

FIG. 9.8   The probabilities for each model: forward scan.

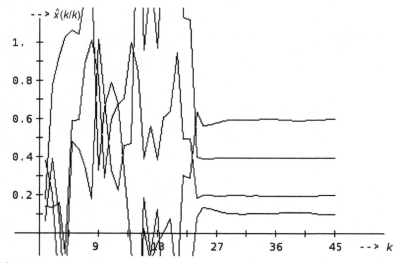

FIG. 9.9   The mean state estimate: forward scan.

shown. For the final mean state estimate, the values of the concentrations involved should be 0.2, 0.4, and 0.6 and the baseline 0.1. Both figures demonstrate that after about 25 measurements the system has converged to the one right model with its corresponding state estimate.

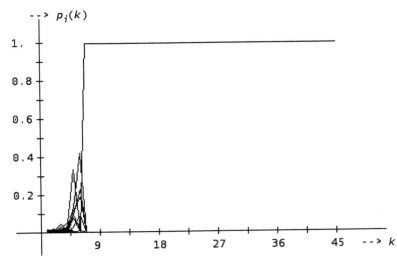

FIG. 9.10 The probabilities for each model: online optimal selection.

In Figs. 9.10 and 9.11, similar results are shown for the online optimal selection of wavelengths processing the measurements. Here, the right model and its estimated state are already found within eight measurements.

Examination of the online means and chi-square values based on the standardized innovations indicates in both cases that the determined best model is indeed a correct and valid one.

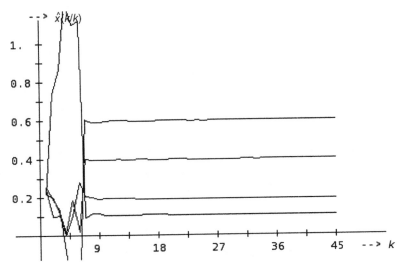

FIG. 9.11 The mean state estimate: online optimal selection.

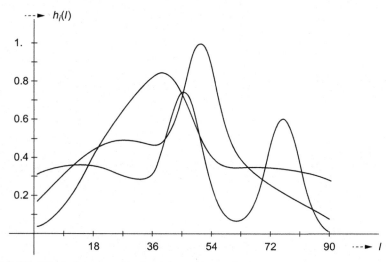

FIG. 9.12 Principal components: the molar spectra used.

The state space model (9.40) can be extended to include stochastic drift.

Example 9.3: This example of principal component analysis was taken from chromatography followed by spectroscopy. Consider a three-component system, with the molar spectra of the pure components plotted in Fig. 9.12 and their associated asymmetric concentration profiles are in Fig. 9.13. Combining them both and adding Gaussian white noise with zero mean and a variance 10^{-5}

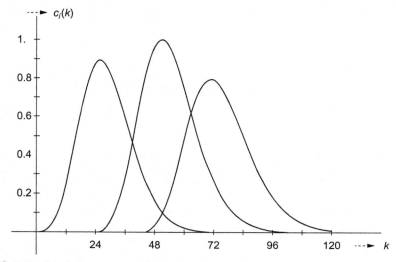

FIG. 9.13 Principal components: the concentrations used.

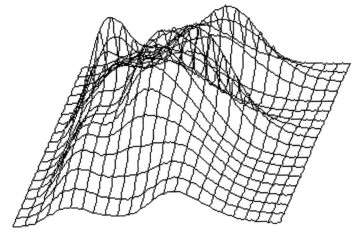

FIG. 9.14   A selection of the resulting measurements.

generate the measurements. The measurements comprise a 90*120 dimensional matrix, i.e., 90 wavelengths and 120 times. The three-dimensional plot in Fig. 9.14 depicts a selection of the measurements employed. It is assumed in advance that five eigenvectors are satisfactory to model the chromatographic system. Forward scanning through the measurements and applying the updating Eqs. (9.16), (9.17), (9.12), and (9.13) for $n = 5$ result in Fig. 9.15 for

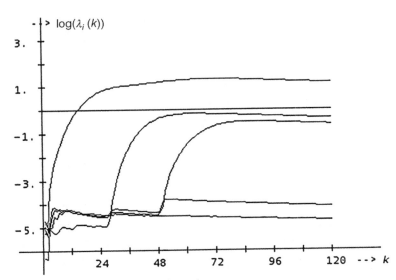

FIG. 9.15   Online eigenvalues: forward scanning.

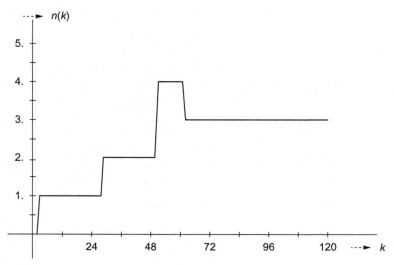

FIG. 9.16   Online number of components: forward scan.

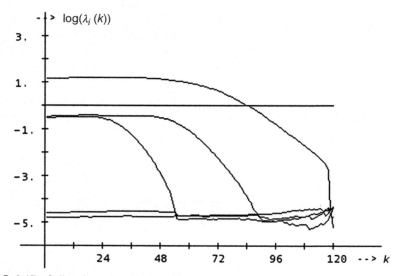

FIG. 9.17   Online eigenvalues: backward scanning.

the estimated eigenvalues. In Fig. 9.16, the evaluated number of significant principal components by Eq. (9.18) is plotted with a critical 99% confidence level. The same is done for backward scanning the measurements resulting in Figs. 9.17 and 9.18. The figures demonstrate that the eigenvalues follow online the concentration profiles hidden in the measurements. From here, the number of

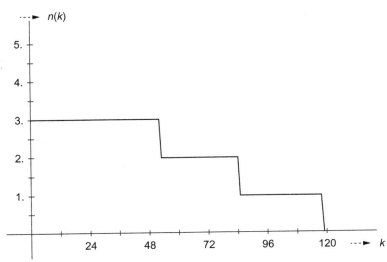

FIG. 9.18   Online number of components: backward scan.

significant principal components can be evaluated online. The final number of significant principal components for both the forward and backward scan is found to be three as should be. The information of the forward and backward scan combined indicates where the chromatographic peaks are located.

The estimated eigenvectors at the end of the forward scan are plotted in Figs. 9.19 and 9.20 for the wavelength-axis (row mode) and time-axis (column mode). The eigenvectors are scaled according to their largest minimum or maximum.

The principal components can also be estimated by Eqs. (9.10)–(9.14). This algorithm produces better results but is about 14 times slower in computation time for the example considered. In the beginning, deviations in the evaluated number of significant principal components occur for the alternative algorithm. Graphics for the alternative algorithm shall not be given because they are quite similar to the example depicted. The local iteration for modeling the principal components by Eqs. (9.20) and (9.21) can be employed offline afterward.

Example 9.4: One result from iterative target transformation factor analysis should be mentioned. This type of factor analysis can only be done offline afterward.

The same chromatographic system as Example 9.3 is employed. Here, valuable information is derived from the measurements, i.e., the final number of significant principal components and the positions where the concentration peaks are located.

The number of concentration peaks in the chromatographic system equals the number of p significant principal components. The final estimated

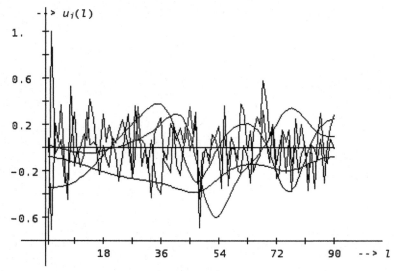

FIG. 9.19   Final eigenvectors forward scan: wavelength-axis (row mode).

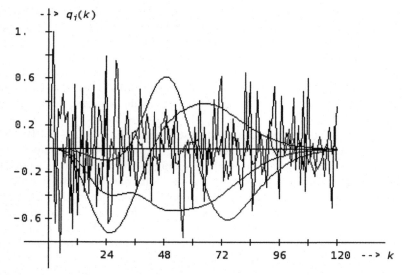

FIG. 9.20   Final eigenvectors forward scan: time-axis (column mode).

eigenvector matrices are redimensioned according to the $m*p$ matrix $U(k)$ and $k*p$ matrix $Q(k)$.

Each concentration peak is a linear combination of the estimated eigenvectors $\mathbf{q}_i(k)$; $i = 1,2,...,p$ contained in the eigenvector matrix $Q(k)$. Using this

gives the p-column coefficient vector $\mathbf{b}_i$ and k-column concentration vector $\mathbf{c}_i$
as:

$$\mathbf{b}_i = Q^{\mathrm{T}}(k)\mathbf{c}_i \tag{9.41}$$

$$\mathbf{c}_i = Q(k)\mathbf{b}_i$$

Now, the computed concentrations contained in $\mathbf{c}_i$ are restricted to nonnegative values and there is only one maximum allowed. Negative concentrations are set to zero and possible secondary maxima are removed. A peak with a triangular shape initializes the concentration profile. The initial and computed concentration vectors $\mathbf{c}_i$ are divided by their Euclidian norm. By iterating through Eq. (9.41) and corresponding restrictions, the computed concentrations converge to the real relative concentration peak until this ends somewhere. Hereafter, the estimated $\mathbf{b}_i$ and $\mathbf{c}_i$ are collected in the $p*p$ matrix B and $k*p$ matrix C, respectively. For each available concentration peak, the described iteration has to be done again.

The m-column vectors $\mathbf{t}_i(k)$ are calculated from the estimated eigenvectors $\mathbf{u}_i(k)$ and eigenvalues $\lambda_i(k)$ for $i = 1,2,...,p$:

$$\mathbf{t}_i(k) = \mathbf{u}_i(k)\sqrt{k\lambda_i(k)} \tag{9.42}$$

The vectors $\mathbf{t}_i(k)$; $i = 1,2,...,p$ are collected in the $m*p$ matrix $T(k)$. This gives finally the $m*p$ matrix A containing the evaluated individual relative spectra:

$$A = T(k)B^{-\mathrm{T}} \tag{9.43}$$

From Example 9.3 follows the final number of significant principal components $p = 3$ and the positions where the concentration peaks are located. The results from iterative target transformation factor analysis are depicted in Figs. 9.21 and 9.22 for the relative concentration peaks and individual relative spectra, respectively. Both are corrected according to their maximum value. In Fig. 9.21, the real relative concentration peaks and the progress of the iterated evaluated relative concentration peaks are shown. Fig. 9.22 shows an almost perfect match of the individual relative molar spectra compared with the evaluated relative spectra.

Example 9.5: The generalized standard addition method (GSAM) followed by spectroscopy is applied for a one-component system with five sensors. The initial volume of the sample is $V_0 = 25\,\mathrm{mL}$ and its initial concentration 0.001 M. Nineteen volume additions of 1 mL are done with a titrant concentration of 0.01 M.

In Fig. 9.23, the measurements are shown and in Fig. 9.24 the corrected measurements are depicted. The measurement noise variance R applied is 10^{-5}. The noise variance in the corrected measurements can be found by the relation $\bar{v}^2 R$, where $\bar{v} = V_0 + 1/(n+1)\sum_{i=0}^{n}\Delta v_i$ is the mean of all the volumes involved, V_0 the initial volume of the sample, and n the number of volume additions Δv_i.

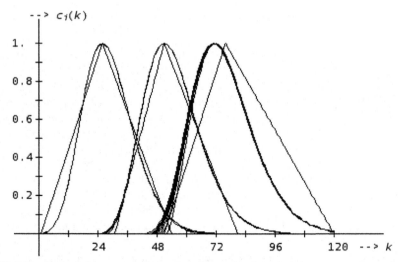

FIG. 9.21   Real relative concentration peaks and iterated evaluated relative concentration peaks by iterative target transformation factor analysis.

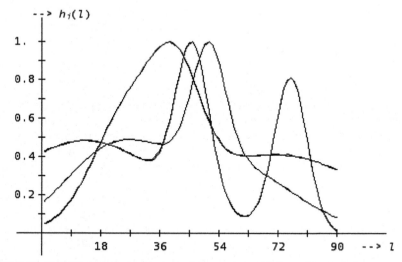

FIG. 9.22   Individual relative molar spectra in comparison with the evaluated relative spectra by iterative target transformation factor analysis.

This gives for this example $\bar{v} = 34.5\,\mathrm{mL}$ and thus a noise variance $1.190 \cdot 10^{-2}$ in the corrected measurements has to be used.

In Fig. 9.25, the online estimated sensitivities by recursive least squares or the Kalman filter are shown for five sensors. The values of the simulated and

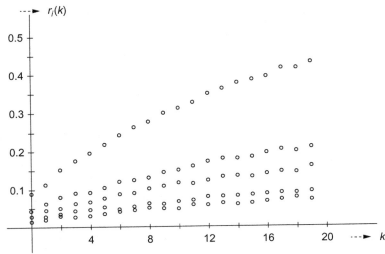

FIG. 9.23 The generalized standard addition method for one component: the measurements for five sensors.

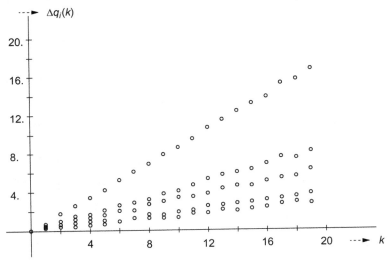

FIG. 9.24 The corrected measurements.

estimated sensitivities can be found in Table 9.1, where also the initial concentration in the sample evaluated offline afterward is given. Note that the estimated sensitivities and the initial concentration in the sample of the classical approach and the final results of the recursive estimation are similar as should be.

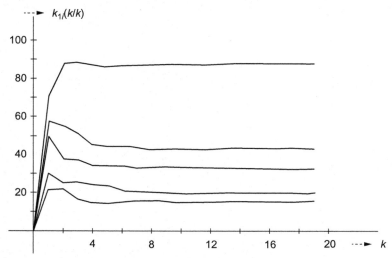

FIG. 9.25   The estimated sensitivities by the Kalman filter.

Finally, the online means and chi-square values of the standardized innovations are depicted in Figs. 9.26 and 9.27, respectively. It shows that the values computed for the means give a good match between its lower and upper 99% confidence limits and that the chi-square values never exceed the upper 99% confidence limit.

Example 9.6: Now, the generalized standard addition method (GSAM) followed by spectroscopy is applied for a three-component system with five sensors.

The sample is divided into three aliquots each with an initial volume $V_0 = 25$ mL and initial concentrations of the three unknown components 0.001 M. Nineteen volume additions of component one of 1 mL are done to aliquot 1 with a titrant concentration 0.01 M. Thereafter, follows 19 volume additions of 1 mL of component two to aliquot 2 and 19 volume additions of 1 mL of component three to aliquot 3, both with the same titrant concentration 0.01 M.

In Fig. 9.28 the measurements are shown and in Fig. 9.29 the corrected measurements are depicted. The measurement noise variance R applied is 10^{-5}. For this example, the mean volume is also $\bar{v} = 34.5$ mL and a noise variance $1.190 \cdot 10^{-2}$ in the corrected measurements has to be used (see Example 9.5).

In Fig. 9.30, the online estimated sensitivities by recursive least squares or the Kalman filter are shown for the three components with five sensors. Note that the applied experimental design allows the use of 15 one-dimensional Kalman filters. The values of the simulated and estimated sensitivities can further be found in Table 9.1. Here, also the three initial concentrations in the sample evaluated offline afterward are given. Note that the estimated sensitivities and the three initial concentrations in the sample of the classical least squares approach and

TABLE 9.1 Generalized standard addition method with the estimated sensitivities for five sensors and initial concentration(s) in the sample for one and three components

Example 9.5

	k_{11}	k_{12}	k_{13}	k_{14}	k_{15}	C_{01}
Simulated	31.99	88.86	42.98	19.59	15.35	0.001
Classical	32.59	88.36	43.30	20.00	15.53	9.973E−4
Recursive ($k=19$)	32.59	88.36	43.30	20.00	15.53	9.973E−4

Example 9.6

	k_{11}	k_{12}	k_{13}	k_{14}	k_{15}	C_{01}	C_{02}	C_{03}
Simulated	31.99	88.86	42.98	19.59	15.35	0.001	0.001	0.001
Classical	32.40	88.92	42.84	18.95	15.63	9.716E−4	9.804E−4	1.019E−3
Recursive ($k=19$)	32.40	88.92	42.84	18.95	15.63	9.716E−4	9.804E−4	1.019E−3
	k_{21}	k_{22}	k_{23}	k_{24}	k_{25}	C_{01}	C_{02}	C_{03}
Simulated	77.04	17.32	24.14	32.03	46.24	0.001	0.001	0.001
Classical	77.46	17.39	25.06	33.07	46.65	1.018E−3	1.018E−3	9.891E−4
Recursive ($k=19$)	77.46	17.39	25.06	33.07	45.65	1.018E−3	1.018E−3	9.891E−4
	k_{31}	k_{32}	k_{33}	k_{34}	k_{35}	C_{01}	C_{02}	C_{03}
Simulated	50.56	35.03	65.45	80.03	25.23	0.001	0.001	0.001
Classical	48.69	35.11	66.48	79.66	25.50	9.842E−4	1.026E−3	9.707E−4
Recursive ($k=19$)	48.69	35.11	66.48	79.66	25.50	9.842E−4	1.026E−3	9.707E−4

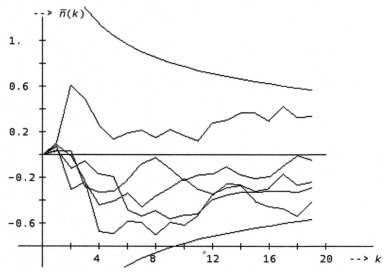

FIG. 9.26 The online means of the standardized innovations with their 99% confidence limits.

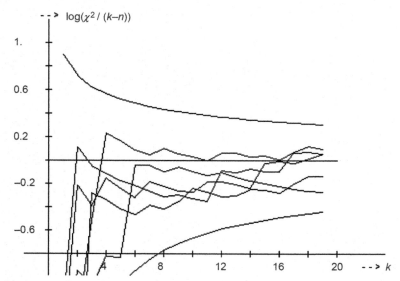

FIG. 9.27 The online chi-square values of the standardized innovations with their 99% confidence limits.

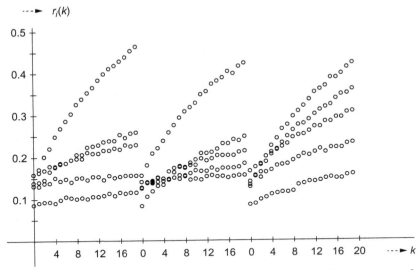

FIG. 9.28 The generalized standard addition method for three components: the measurements for five sensors.

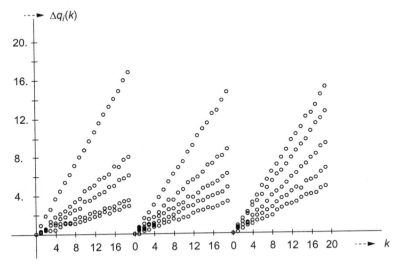

FIG. 9.29 The corrected measurements.

the final results of the recursive estimation are similar as should be. At last, the online means and chi-square values of the standardized innovations are depicted in Figs. 9.31 and 9.32. The results are the same as for Example 9.5.

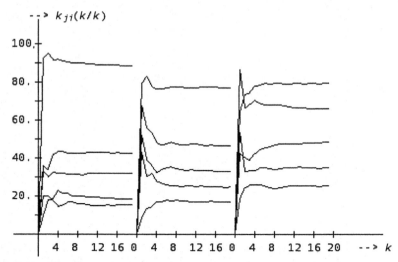

FIG. 9.30   The estimated sensitivities by the Kalman filter for the three components.

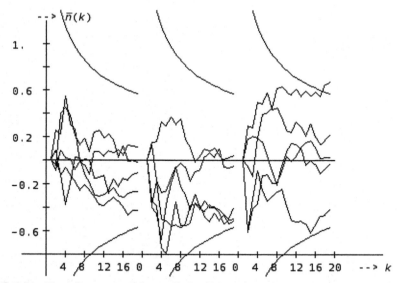

FIG. 9.31   The online means of the standardized innovations with their 99% confidence limits.

Example 9.7: Now, the generalized standard addition method (GSAM) will be adjusted for the same Example 9.6 followed by spectroscopy. There are problems with both verification algorithms for the online means and chi-square values based on the standardized innovations. Because of the use of difference measurements, there is no longer Gaussian white noise involved anymore and a

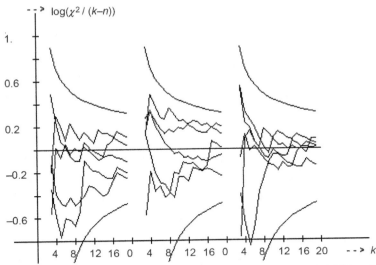

FIG. 9.32   The online chi-square values of the standardized innovations with their 99% confidence limits.

revision of the model used has to be applied. Both the means and chi-square values exceed the 99% confidence limits most of the times when they are used in different simulations of the system and the given results in Examples 9.5 and 9.6 are just a lucky shot.

The measurement model applied for one sensor of both classical least squares and the Kalman filter is given by an $(r + 1)$-dimensional state space model with the added baseline b_i:

$$q_{ki} = r_{ki}(V_0 + \Delta v_k) = \left(\Delta \mathbf{n}_k^{\mathrm{T}} \ 1 \right) \begin{pmatrix} \mathbf{k}_i \\ b_i \end{pmatrix} + e_i \tag{9.44}$$

With this state space model, the Gaussian whiteness of the standardized innovations is preserved. Another advantage of the proposed model is that the estimated baseline is based upon more measurements instead of subtracting one measurement at $k = 0$ to get difference values. Note that here the applied experimental design allows the use of 15 two-dimensional Kalman filters, i.e., three components times five sensors. Classical least squares based on Eq. (9.44) can be implemented easily and is performed three times for each component separately and done for five sensors simultaneously.

In Fig. 9.33 the uncorrected measurements are shown and in Fig. 9.34 the volume corrected measurements without difference values are depicted. In Figs. 9.35 and 9.36, the online estimated sensitivities and baselines by recursive least squares or the Kalman filter are shown for the three components with five sensors. The estimated sensitivities $\mathbf{k}_i$ and baselines b_i are collected in the

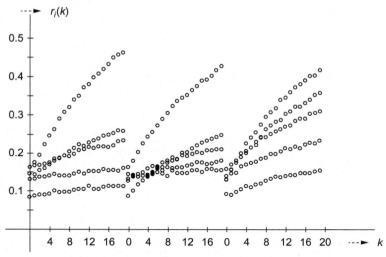

FIG. 9.33 The revised generalized standard addition method for three components: the measurements for five sensors.

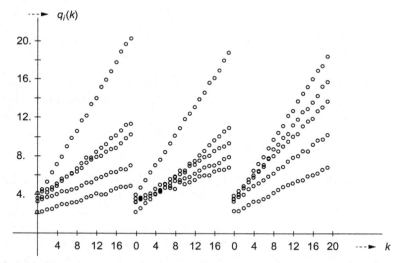

FIG. 9.34 The corrected measurements.

matrix K and vector **b**, respectively. The values of the simulated and estimated sensitivities can further be found in Table 9.2. Here, also the three initial concentrations in the sample evaluated offline afterward are given.

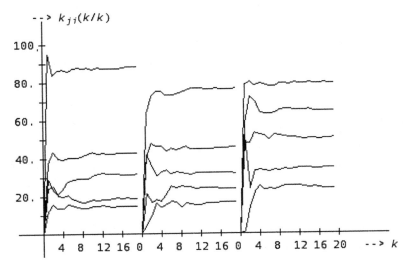

FIG. 9.35   The estimated sensitivities by the Kalman filter for the three components.

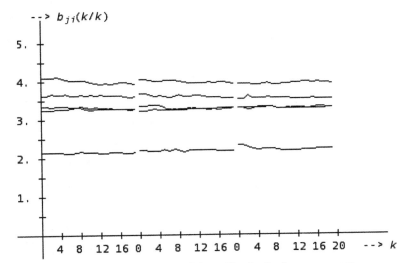

FIG. 9.36   The estimated baselines by the Kalman filter for the three components.

The initial concentrations in the sample are evaluated by:

$$\mathbf{n}_0 = \left(KK^{\mathrm{T}}\right)^{-1} K\mathbf{b} \tag{9.45}$$

$$\mathbf{c}_0 = \mathbf{n}_0 / V_0$$

Note that the estimated sensitivities and the three initial concentrations in the sample of the classical least squares approach and the final results of the

TABLE 9.2 Revised generalized standard addition method with the estimated sensitivities for five sensors and initial concentration(s) in the sample for three components

Example 9.7	k_{11}	k_{12}	k_{13}	k_{14}	k_{15}	c_{01}	c_{02}	c_{03}
Simulated	31.99	88.86	42.98	19.59	15.35	0.001	0.001	0.001
Classical	32.32	88.73	43.17	19.47	15.16	1.037E-3	9.734E-4	9.853E-4
Recursive ($k=19$)	32.32	88.73	43.17	19.47	15.16	1.037E-3	9.734E-4	9.853E-4
	k_{21}	k_{22}	k_{23}	k_{24}	k_{25}	c_{01}	c_{02}	c_{03}
Simulated	77.04	17.32	24.14	32.03	46.24	0.001	0.001	0.001
Classical	77.25	17.33	24.30	32.29	46.06	1.015E-3	9.738E-4	1.007E-3
Recursive ($k=19$)	77.25	17.33	24.30	32.29	46.06	1.015E-3	9.738E-4	1.007E-3
	k_{31}	k_{32}	k_{33}	k_{34}	k_{35}	c_{01}	c_{02}	c_{03}
Simulated	50.56	35.03	65.45	80.03	25.23	0.001	0.001	0.001
Classical	51.03	35.16	65.47	79.66	24.41	9.934E-4	1.001E-3	1.005E-3
Recursive ($k=19$)	51.03	35.16	65.47	79.66	24.41	9.934E-4	1.001E-3	1.005E-3

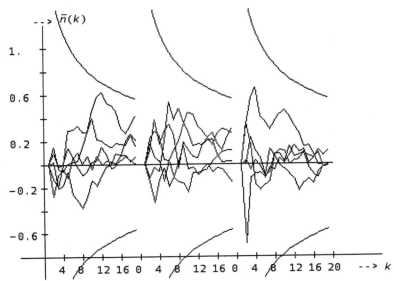

FIG. 9.37   The online means of the standardized innovations with their 99% confidence limits.

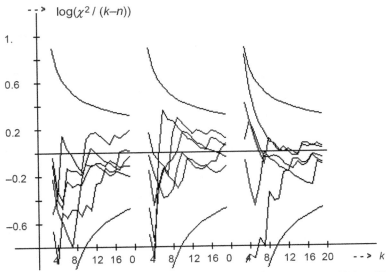

FIG. 9.38   The online chi-square values of the standardized innovations with their 99% confidence limits.

recursive estimation are similar as should be. Finally, the online means and chi-square values of the standardized innovations are depicted in Figs. 9.37 and 9.38. These results are overall better in comparison with the same ones in Example 9.6.

Chapter 10

Subspace identification methods

Chapter outline

10.1 Subspace identification methods

Subspace identification methods are used to identify the parameters of state space models from the corresponding input/output data. These methods have their origin in state space realization theory as developed in the 60s, and were firstly discussed by Ho and Kalman (1966), who proposed a method for recovering the system matrices from the system's impulse responses. The main drawback on using realization theory for system identification is that a reliable nonparametric estimate of the impulse response is difficult to obtain. This is especially true when dealing with the dynamics associated with chemical or biochemical reactions as analyzed in a chemometrics context. Furthermore, special excitatory inputs may also need to be used such as a Dirac impulse, a pseudorandom binary sequence or other types of white noise sequences. Such signals, however, may be difficult to be realized in many experimental settings. Extracting the desired information from input/output data directly without explicitly requiring the impulse response is one of the main advantages of subspace identification methods. In this sense, this approach is complementary to existing Kalman filter (KF) approaches.

A more careful look of the name of these methods implies that the identified model parameters for a linear system are obtained using a combination of rows and columns of matrices (subspaces) generated from the input/output datasets. In order to clarify the general ideas behind subspace identification, one has to invoke ideas from projective geometry. In two dimensions, the rows of any matrix A can be projected into a matrix B and its orthogonal counterpart $B^{\perp}$ as shown in the Fig. 10.1A below. One should visualize this procedure as a simple rotation operator, after asking: by how much do I need to multiply my vector to get the right value for each projection. In our case, however, we deal with

State Estimation in Chemometrics. https://doi.org/10.1016/B978-0-08-102603-8.00010-9

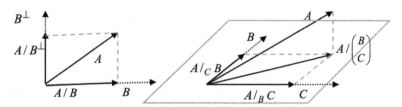

FIG. 10.1 (A) orthogonal projection of matrix A to matrices B and $B^\perp$ and (B) oblique projection of the row space A to the row space of $\begin{pmatrix} B \\ C \end{pmatrix}$.

matrices so we do that for every column of the matrix. A numerically efficient and robust computation of the orthogonal projection can be done by an LQ decomposition:

$$\begin{pmatrix} B \\ A \end{pmatrix} = LQ = \begin{pmatrix} L_{11} & 0 \\ L_{21} & L_{22} \end{pmatrix} \begin{pmatrix} Q_1 \\ Q_2 \end{pmatrix} \tag{10.1}$$

so:

$$A/B = L_{21}Q_1, \quad A/B^\perp = L_{22}Q_2$$

In a similar manner, the oblique projection of the row space of A along the row space of B on the row space of C as shown in Fig. 10.1B is defined as:

$$A/_B C = A(C^T \ B^T) \left[\left(\begin{pmatrix} C \\ B \end{pmatrix} (C^T \ B^T) \right)^\dagger \right] C \tag{10.2}$$

This result can be derived from an orthogonal projection of the row space A to the row space of $\begin{pmatrix} B \\ C \end{pmatrix}$.

An effective way of achieving this is using the LQ decomposition again:

$$\begin{pmatrix} B \\ C \\ A \end{pmatrix} = \begin{pmatrix} L_{11} & 0 & 0 \\ L_{21} & L_{22} & 0 \\ L_{31} & L_{32} & L_{33} \end{pmatrix} \begin{pmatrix} Q_1 \\ Q_2 \\ Q_3 \end{pmatrix} \tag{10.3}$$

Then, the orthogonal projections can be written as:

$$A/_B C = L_{32}L_{22}^{-1}C = L_{32}L_{22}^{-1}(L_{21} \ L_{22}) \begin{pmatrix} Q_1 \\ Q_2 \end{pmatrix} \tag{10.4}$$

The oblique projection of the row space of A onto the row space of C along the row space of B is defined as:

$$A/_B C = [A/B^\perp][C/B^\perp]^\dagger C = A\Pi_{B^\perp} + (C\Pi_{B^\perp})^\dagger C \tag{10.5}$$

The oblique projection essentially decomposes the matrix A into three matrices:

$$A = A/_B C + A/_C B + A/\begin{pmatrix} B \\ C \end{pmatrix}, \tag{10.6}$$

where the first term is the oblique projection of the row space of A onto the row space of C along the row space of B, the second term is the oblique projection of the row space of A onto the row space of B along the row space of C, and the last term denotes the projection of matrix A on the BC plane.

$$\begin{bmatrix} U_p \\ \overline{U_f} \end{bmatrix} = \begin{bmatrix} u(0) & u(1) & \cdots & u(j-1) \\ u(1) & u(2) & \cdots & u(j) \\ \vdots & \vdots & \vdots & \vdots \\ u(i-1) & u(i) & \cdots & u(i+j-2) \\ \hline u(i) & u(i+1) & \cdots & u(i+j-1) \\ u(i+1) & u(i+2) & \cdots & u(i+j) \\ \vdots & \vdots & \vdots & \vdots \\ u(i+h-1) & u(i+h) & \cdots & u(i+h+j-2) \end{bmatrix}, \tag{10.7}$$

$$\begin{bmatrix} U_p^+ \\ \overline{U_p^-} \end{bmatrix} = \begin{bmatrix} u(0) & u(1) & \cdots & u(j-1) \\ u(1) & u(2) & \cdots & u(j) \\ \vdots & \vdots & \vdots & \vdots \\ u(i-1) & u(i) & \cdots & u(i+j-2) \\ u(i) & u(i+1) & \cdots & u(i+j-1) \\ \hline u(i+1) & u(i+2) & \cdots & u(i+j) \\ \vdots & \vdots & \vdots & \vdots \\ u(i+h-1) & u(i+h) & \cdots & u(i+h+j-2) \end{bmatrix}$$

$$\begin{bmatrix} Y_p \\ \overline{Y_f} \end{bmatrix} = \begin{bmatrix} y(0) & y(1) & \cdots & y(j-1) \\ y(1) & y(2) & \cdots & y(j) \\ \vdots & \vdots & \vdots & \vdots \\ y(i-1) & y(i) & \cdots & y(i+j-2) \\ \hline y(i) & y(i+1) & \cdots & y(i+j-1) \\ y(i+1) & y(i+2) & \cdots & y(i+j) \\ \vdots & \vdots & \vdots & \vdots \\ y(i+h-1) & y(i+h) & \cdots & y(i+h+j-2) \end{bmatrix}, \tag{10.8}$$

$$\begin{bmatrix} Y_p^+ \\ Y_p^- \end{bmatrix} = \begin{bmatrix} y(0) & y(1) & \cdots & y(j-1) \\ y(1) & y(2) & \cdots & y(j) \\ \vdots & \vdots & \vdots & \vdots \\ y(i-1) & y(i) & \cdots & y(i+j-2) \\ y(i) & y(i+1) & \cdots & y(i+j-1) \\ \hline y(i+1) & y(i+2) & \cdots & y(i+j) \\ \vdots & \vdots & \vdots & \vdots \\ y(i+h-1) & y(i+h) & \cdots & y(i+h+j-2) \end{bmatrix}$$

One may thus assume a linear dynamic system with deterministic and stochastic inputs described in standard state space innovation form:

$$x(k+1) = Ax(k) + Bu(k) + Ke(k) \tag{10.9}$$

$$y(k) = Cx(k) + Du(k) + e(k), \tag{10.10}$$

where K denotes the stationary Kalman gain. In subspace system identification the input and output data matrices are arranged as input U_p, U_f and output Y_p, Y_f Hankel matrices (with subscripts p and f denoting past and future states). In Eqs. (10.7), (10.8) j is the number of data points, i is a user-defined index such that $n < i < j$ and n is the system order.

Matrices U_p^+, U_p^- are created from U_p, U_f by moving the first block row from U_f to the end of U_p. This variation is later used to retrieve the system matrices. For the use in 4SID algorithms, all signals (inputs, outputs and noises) are arranged into the Hankel matrices.

For the outputs $y(k)$ and the noises $e(k)$ similar Hankel matrices Y_p, Y_f, E_p, E_f can be constructed. A combination of U_p, U_f denoted as $W_p = \begin{pmatrix} U_p \\ Y_p \end{pmatrix}$ is used as a regressor. System-state sequence is also used in matrix form with the following structure:

$$\begin{aligned} X_p &= (x(0) \quad x(1) \quad \ldots \quad x(j-1)) \\ X_f &= (x(i) \quad x(i+1) \quad \ldots \quad x(i+j-1)) \end{aligned} \tag{10.11}$$

One of the principal ideas in 4SID methods is to combine the recursive state space model into a single linear matrix equation, relating the signal matrices with the parameters matrices. For the use in the 4SID algorithms, all signals (inputs, outputs, and noises) are arranged into the Hankel matrices. One should assume a known set of input/output data samples.

There are also parameter-related matrices: The extended observability matrix Γ_k is an extension of observability matrix for a number of block rows higher than the system order $k \geq n$:

$$\Gamma_k = \begin{pmatrix} C \\ CA \\ \vdots \\ CA^{k-1} \end{pmatrix} \qquad (10.12)$$

Similarly, the reversed extended controllability matrices Δ_k^d, Δ_k^s corresponding to the deterministic (d) and stochastic (s) parts, respectively, are defined as:

$$\Delta_k^d = \begin{pmatrix} A^{k-1}B & A^{k-2}B & \cdots & B \end{pmatrix} \qquad (10.13)$$

$$\Delta_k^s = \begin{pmatrix} A^{k-1}K & A^{k-2}K & \cdots & K \end{pmatrix} \qquad (10.14)$$

The last two matrices H_k^d, H_k^s are Toeplitz matrices composed from the impulse responses (Markov parameters) of deterministic and stochastic subsystems, respectively.

$$H_k^d = \begin{pmatrix} D & 0 & 0 & \cdots & 0 \\ CB & D & 0 & \cdots & 0 \\ CAB & CB & D & \cdots & 0 \\ \vdots & \vdots & \vdots & \ddots & \vdots \\ CA^{k-2}B & CA^{k-3}B & CA^{k-4}B & \cdots & D \end{pmatrix}, \qquad (10.15)$$

$$H_k^s = \begin{pmatrix} I & 0 & 0 & \cdots & 0 \\ CK & I & 0 & \cdots & 0 \\ CAK & CK & I & \cdots & 0 \\ \vdots & \vdots & \vdots & \ddots & \vdots \\ CA^{k-2}K & CA^{k-3}K & CA^{k-4}K & \cdots & I \end{pmatrix}$$

The basic problem that needs to be solved can be thus stated as follows: Given s number of samples of the input sequence $\{u(0), \cdots u(s-1)\}$ and the output sequence $\{y(0), \cdots y(s-1)\}$, estimate the parameters of the combined deterministic-stochastic model. Essentially this argument implies that one needs to estimate the system order n and to obtain system matrices A, B, C, D, K and covariance matrix R_e of the noise $e(k)$. The inputs, the outputs, and the states are known and they are related by the state space model, which can be written into a set of linear equations:

$$\begin{pmatrix} x(i+1) & x(i+2) & \cdots & x(i+j-1) \\ y(i) & y(i+1) & \cdots & y(i+j-2) \end{pmatrix}$$
$$= \begin{pmatrix} A & B \\ C & D \end{pmatrix} \begin{pmatrix} x(i) & x(i+1) & \cdots & x(i+j-2) \\ u(i) & u(i+1) & \cdots & u(i+j-2) \end{pmatrix} \quad (10.16)$$

and solved using least squares or total least squares to get the model parameters A, B, C, D.

As already mentioned before, the starting point of 4SID methods is a combination of the recursive state space innovation model into one single linear matrix equation. This can be done with Hankel matrices by recursively substituting $x(k+1) = Ax(k) + Bu(k) + Ke(k)$ into $y(k) = Cx(k) + Du(k) + e(k)$ using:

$$Y_p = \Gamma_i X_p + H_i^d U_p + H_i^s E_p \quad (10.17)$$

$$Y_f = \Gamma_h X_f + H_h^d U_f + H_h^s E_f \quad (10.18)$$

$$X_f = A^i X_p + \Delta_i^d U_p + \Delta_i^s E_p \quad (10.19)$$

The first two equations are similarly defining outputs as a linear combination of previous states by the extended observability matrix (response from the states) and a linear combination of previous inputs and noises by their respective impulses responses H^d, H^s. The last equation is relating the future and the past states under the influence of the inputs and the noises.

In subspace identification, the aim is to estimate invariant subspaces of the extended observability matrix Γ_h and the state sequence matrix X_f from the input/output data. This estimation is also related to the determination of the system order. The important observation is that in order to obtain these subspaces, only the term X_f is needed and its estimate can be obtained from the available data through the projection operations. This term is usually denoted as a matrix O_h given from $O_h = \Gamma_h X_f$ and it can be split into the required subspaces by singular value decomposition (SVD). Description of the matrix O_h content is usually avoided, but each column can be seen as a response of the system to the nonzero initial state from an appropriate column of the matrix X_f, without any deterministic or stochastic inputs.

For a pure deterministic chemical system, the matrix O_h has the rank equal to the system order with the following SVD factorization:

$$O_h = USV^T = (U_1 \ U_2) \begin{pmatrix} S_1 & 0 \\ 0 & 0 \end{pmatrix} \begin{pmatrix} V_1^T \\ V_2^T \end{pmatrix}, \quad (10.20)$$

where S_1 is $n \times n$ submatrix of S containing nonzero singular values of O_h and U_1, V_1^T are the appropriate parts of the matrices U, V^T. The number of nonzero singular values is equal to the order of the system and the required subspaces can be obtained as:

$$\Gamma_h = U_1 S_1^{1/2}, \quad X_f = S_1^{1/2} V_1^T, \tag{10.21}$$

where the square root of S_1 is used simply because this is a diagonal matrix. At this point, one can notice that multiplication by $S_1^{1/2}$ can be omitted, because we are interested only in the spanned spaces. This term is included only for the purpose that the equality $O_h = \Gamma_h X_f$ is true.

In the presence of the noise, the matrix O_h will of course be full rank. Thus all singular values of O_h will be nonzero, i.e., the diagonal of the matrix S will have nonzero entries in nonincreasing order. The rank of the identified chemical system has to be chosen from the number of significant singular values. This can be a rather tricky task, for which few theoretical guidelines are available. Assuming that the task here is that the chemical system order is to be determined, the SVD matrices are partitioned into the "signal" and "noise" parts:

$$O_h = USV^T = (U_s \quad U_n) \begin{pmatrix} S_s & 0 \\ 0 & S_n \end{pmatrix} \begin{pmatrix} V_s^T \\ V_n^T \end{pmatrix}, \tag{10.22}$$

where U_s, V_s^T contain n principal singular vectors, whose corresponding singular values are collected in the $n \times n$ diagonal matrix S_s. The "cleaned" estimates of the extended observability matrix and state sequence matrix are then:

$$\Gamma_h = U_s S_s^{1/2}, \quad X_f = S_s^{1/2} V_s^T \tag{10.23}$$

In subspace identification, the term $\Gamma_h \hat{X}_f$ will be obtained from the input/output data in the signal matrices by an oblique projection. We use $Y_f = \Gamma_h X_f + H_h^d U_f + H_h^s E_f$, where the last two terms on the right side will be eliminated by a projection. This will be used as an approximation of $O_h \doteq \Gamma_h \hat{X}_f$, where $\hat{X}_f$ is the Kalman filter estimate of X_f from the available past data W_p.

Let's try an orthogonal projection of the future output Y_f onto the subspace of past data W_p and future inputs U_f:

$$Y_f / \begin{pmatrix} W_p \\ U_f \end{pmatrix} = \Gamma_h \hat{X}_f + H_h^d U_f \tag{10.24}$$

The orthogonal projection eliminated the noise term, because the estimated state sequence $\hat{X}_f$ and the future inputs U_f lie in the joint row space of W_p and U_f, but the future noise E_f is perpendicular to this subspace when the number of samples increases toward infinity. The reasoning for this becomes clearer if one writes:

$$X_p = \Gamma_i^\dagger Y_p - \Gamma_i^\dagger H_i^d U_p - \Gamma_i^\dagger H_i^s E_p$$

$$= \left(\Gamma_i^\dagger - \Gamma_i^\dagger H_i^d - \Gamma_i^\dagger H_i^s \right) \begin{pmatrix} Y_p \\ U_p \\ E_p \end{pmatrix} \tag{10.25}$$

Substituting this into Eq. (10.19) provides the future states X_f:

$$X_f = A^i X_p + \Delta_i^d U_p + \Delta_i^s E_p$$

$$= A^i \Gamma_i^\dagger Y_p - A^i \Gamma_i^\dagger H_i^d U_p - A^i \Gamma_i^\dagger H_i^s E_p + \Delta_i^d U_p + \Delta_i^s E_p$$

$$= \left(A^i \Gamma_i^\dagger \quad \left(\Delta_i^d - A^i \Gamma_i^\dagger H_i^d \right) \quad \left(\Delta_i^s - A^i \Gamma_i^\dagger H_i^s \right) \right) \begin{pmatrix} Y_p \\ U_p \\ E_p \end{pmatrix} \qquad (10.26)$$

These two last equations clearly indicate that both past states X_p and future states X_f can be obtained as a linear combination of past data Y_p, U_p, E_p. In other words, they lie in their joint row space. To compute an estimate of X_f, the noise term can be replaced by its mean value:

$$\hat{X}_f = \left(A^i \Gamma_i^\dagger \quad \left(\Delta_i^d - A^i \Gamma_i^\dagger H_i^d \right) \right) \begin{pmatrix} Y_p \\ U_p \end{pmatrix} = L_w W_p \qquad (10.27)$$

showing $\hat{X}_f$ to lie in the row space of W_p. The orthogonal projection helps us to get rid of the noise term, but we still need to eliminate also the influence of the future inputs U_f and that is where an oblique projection is the right tool:

$$Y_f /_{U_f} W_p = \Gamma_h \hat{X}_f \qquad (10.28)$$

By making the oblique projection of the future outputs Y_f onto the row space of past data W_p along the row space of future inputs U_f will give the term $\Gamma_i \hat{X}_f$ as shown in Fig. 10.2.

This becomes clearer when we substitute Eq. (10.17) to the individual component terms of Y_f in Eq. (10.18):

$$Y_f /_{U_f} W_p = \left(\Gamma_h X_f + H_h^d U_f + H_h^s E_f \right) /_{U_f} W_p$$

$$= \Gamma_h X_f /_{U_f} W_p + H_h^d U_f /_{U_f} W_p + H_h^s E_f /_{U_f} W_p \qquad (10.29)$$

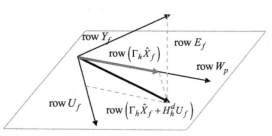

FIG. 10.2 Orthogonal *(black bold)* and oblique *(gray bold)* projections clarifying the subspace identification procedure.

Taking each term separately, when the number of samples and number of past data (rows) is large, $(i, j \to \infty)$:

$$\lim_{i,j\to\infty} X_f /_{U_f} W_p = \lim_{i,j\to\infty} X_f / W_p = \hat{X}_f \qquad (10.30)$$

because $\hat{X}_f$ is the X_f estimate lying in W_p. Furthermore, from the oblique projection definition, $U_f /_{U_f} W_p = 0$. Because noise is independent as the number of samples j is also large we have:

$$\lim_{j\to\infty} E_f /_{U_f} W_p = 0 \qquad (10.31)$$

Therefore,

$$\lim_{i,j\to\infty} Y_f /_{U_f} W_p = \Gamma_h \hat{X}_f \qquad (10.32)$$

From the previous analysis it is possible to assume the inputs, the outputs, and the state sequences are now available for the chemical system under study. By choosing the row space that is the basis for the state sequence:

$$\hat{X}(i) = (\hat{x}(i), \ldots, \hat{x}(i+j-1)), \quad \hat{X}(i+1) = (\hat{x}(i+1), \ldots, \hat{x}(i+j))$$
$$U(i) = (u(i), \ldots, u(i+j-1)), \qquad Y(i) = (y(i), \ldots, y(i+j-1)) \qquad (10.33)$$

in the absence of feedback the parameters of the innovation model can be consistently estimated from the following matrix equation relating all the available data by the state space model:

$$\begin{pmatrix} \hat{X}(i+1) \\ Y(i) \end{pmatrix} = \begin{pmatrix} A & B \\ C & D \end{pmatrix} \begin{pmatrix} \hat{X}(i) \\ U(i) \end{pmatrix} + \varepsilon \qquad (10.34)$$

This solution can be obtained by least squares or total least squares. If:

$$\Theta = \begin{pmatrix} A & B \\ C & D \end{pmatrix}, \quad \Upsilon = \begin{pmatrix} \hat{X}(i+1) \\ Y(i) \end{pmatrix}, \quad \Xi = \begin{pmatrix} \hat{X}(i) \\ U(i) \end{pmatrix} \qquad (10.35)$$

the least squares solution can be obtained as:

$$\Theta = \Upsilon \Xi^\dagger = \Upsilon \Xi^T (\Xi \Xi^T)^{-1} \qquad (10.36)$$

and the associated stochastic parameters can be calculated from:

$$R_e = \Sigma_{22}, \quad K = \Sigma_{12} \Sigma_{22}^{-1}, \qquad (10.37)$$

where

$$\Sigma = \begin{pmatrix} \Sigma_{11} & \Sigma_{12} \\ \Sigma_{21} & \Sigma_{22} \end{pmatrix} = \frac{1}{j-n+m} (\Upsilon - \Theta\Xi)(\Upsilon - \Theta\Xi)^T \qquad (10.38)$$

Again, the consistency of the solution requires $j \to \infty$.

At this point it is also worth noting that it is possible to estimate all the needed parameters from $\mathrm{col}\Gamma_h$. First the extended observability matrix Γ_h is obtained as a basis of $\mathrm{col}\Gamma_h$ or it is obtained directly from the SVD of $\Gamma_h \hat{X}_f$. Then the matrices A, B, C, D are determined in two steps. Firstly the matrix C can be read directly from the first block row of Γ_h. The matrix A is determined from the shift structure of Γ_h. Denoting:

$$\underline{\Gamma}_h = \begin{pmatrix} C \\ \vdots \\ CA^{k-2} \end{pmatrix}, \quad \overline{\Gamma}_h = \begin{pmatrix} CA \\ \vdots \\ CA^{k-1} \end{pmatrix}, \tag{10.39}$$

where $\underline{\Gamma}_h$ is Γ_h without the last block row and $\overline{\Gamma}_h$ is Γ_h without the first block row, the shift structure implies that:

$$\underline{\Gamma}_h A = \overline{\Gamma}_h \tag{10.40}$$

This equation is linear in A and can be solved by least squares or total least squares.

In the second step, determination of matrices B and D is performed. By multiplying the input/output equation $Y_f = \Gamma_h X_f + H_h^d U_f + H_h^s E_f$ from the left by $\Gamma_h^\perp$ and from the right by $U_f^\dagger$ it follows that:

$$\Gamma_h^\perp Y_f U_f^\dagger = \Gamma_h^\perp \Gamma_h X_f U_f^\dagger + \underbrace{\Gamma_h^\perp \, H_h^d U_f \, U_f^\dagger}_{\text{input term}} + \underbrace{\Gamma_h^\perp \, H_h^s E_f \, U_f^\dagger}_{\text{noise term}} \tag{10.41}$$

But since $\Gamma_h^\perp$ is a full row matrix satisfying $\Gamma_h^\perp \Gamma_h = 0$, the equation may be simplified to:

$$\Gamma_h^\perp Y_f U_f^\dagger = \Gamma_h^\perp H_h^d \tag{10.42}$$

By denoting the ith column of the left hand side as M_i and the ith column of $\Gamma_h^\perp$ as L_i then:

$$(M_1 \; M_2 ... M_h) = (L_1 \; L_2 ... L_h) \begin{pmatrix} D & 0 & 0 & \cdots & 0 \\ CB & D & 0 & \cdots & 0 \\ CAB & CB & D & \cdots & 0 \\ \vdots & \vdots & \vdots & \ddots & \vdots \\ CA^{h-2}B & CA^{h-3}B & CA^{h-4}B & \cdots & D \end{pmatrix} \tag{10.43}$$

As discussed in Van Overschee and De Moor (1996), the expression can be rewritten as:

$$\begin{pmatrix} M_1 \\ M_2 \\ \vdots \\ M_h \end{pmatrix} = \begin{pmatrix} L_1 & L_2 & \cdots & L_{h-1} & L_h \\ L_2 & L_3 & \cdots & L_h & 0 \\ L_3 & L_4 & \cdots & 0 & 0 \\ \vdots & \vdots & \ddots & \vdots & \vdots \\ L_h & 0 & 0 & \cdots & 0 \end{pmatrix} \begin{pmatrix} I & 0 \\ 0 & \Gamma_h \end{pmatrix} \begin{pmatrix} D \\ B \end{pmatrix}, \tag{10.44}$$

which is an overdetermined set of linear equations in the unknowns B and D and it can be solved by least squares or total least squares.

At this point, some generic remarks regarding subspace methods may be discussed. The deterministic inputs, disturbances, and response from the states will be shown to generate certain high-dimensional subspaces (spanned by the rows of appropriate Hankel matrices). The output of the chemical LTI system will be shown to generate similar subspaces as a linear combination of these subspaces. In this context, the power of projections lies in their ability to separate this linear combination from the knowledge of the inputs and outputs. The orthogonal projection is used to asymptotically eliminate the influence of the disturbances from the joint space. This can be viewed as a separation of an effective signal from the noise. The more capable oblique projection is also able to simultaneously eliminate the influence of the disturbances and moreover to eliminate the influence of the deterministic measurable signal (the input). This is mainly used in 4SID methods to obtain the subspace generated only by the state sequence, which is later used to obtain the model parameters.

Furthermore, the effect of the projections should be visualized as a necessary transformation that is required to separate different subspaces associated with a combination of input/output signals as well as states, which are joined together through the linear combination of multiple signals. More specifically, the orthogonal projection can be used to asymptotically eliminate the noise in the signal and the oblique projection can be used to eliminate both the influence of the disturbances as well as the influence of the deterministic measurable signal (the input to the system). Through the oblique projection transformations, the system state sequence can be obtained so that the model parameters can be obtained.

The 4SID algorithm can be interpreted as KF, estimating the system states without prior knowledge of the system parameters (matrices). This is possible due to an arbitrary state space basis choice of SS model. The 4SID methods estimate only the row space of state sequence matrix X_f, which is independent of the basis selection and consequently use the fact that any of its basis is a valid state sequence. If an initial state estimate $\hat{x}(0)$, the initial estimate of the matrix $P(0)$ and the input $u(0), \ldots u(k-1)$ and output $y(0), \ldots y(k-1)$ measurements are given, then the non-steady-state Kalman filter state estimate $\hat{x}(k)$ is defined by the following recursive formulas:

$$\hat{x}(k) = A\hat{x}(k+1) + Bu(k) + K(k-1)(y(k-1) - C\hat{x}(k-1) - Du(k-1)) \quad (10.45)$$

$$K(k-1) = \left(G - AP(k-1)C^T\right)\left(\Lambda(0) - CP(k-1)C^T\right)^{-1} \quad (10.46)$$

$$P(k) = AP(k-1)A^T + \left(G - AP(k-1)C^T\right)\left(\Lambda(0) - CP(k-1)C^T\right)^{-1}\left(G - P(k-1)C^T\right)^T$$

$$(10.47)$$

where G is state to output covariance matrix and $\Lambda(0)$ is the output covariance matrix, which are, respectively, defined as:

$$G = A\Sigma C^T + S, \quad \Lambda(0) = C\Sigma C^T + R, \tag{10.48}$$

where Σ is the state covariance matrix $E[x(k)x^T(k)] = \Sigma$ and given by the solution of the Lyapunov equation:

$$\Sigma = A\Sigma A^T + Q \tag{10.49}$$

As discussed in Van Overschee and De Moor (1996), the recursive KF formula presented here can be rewritten into one single linear matrix equation as:

$$\hat{x}(k) = \left(A^k - \Omega_k \Gamma_k \mid \Delta_k^d - \Omega_k H_k^d \mid \Omega_k \right) \begin{pmatrix} \hat{x}(0) \\ \hline u(0) \\ \vdots \\ u(k-1) \\ \hline y(0) \\ \vdots \\ y(k-1) \end{pmatrix}, \tag{10.50}$$

where:

$$\Omega_k = \left(\Delta_k^c - A^k P_0 \Gamma_k^T \right) \left(L_k - \Gamma_k P_0 \Gamma_k^T \right)^{-1} \tag{10.51}$$

The significance of the KF matrix form is that it indicates how the Kalman filter state estimate $\hat{x}(k)$ can be written as a linear combination of the past input $u(0), \ldots u(k-1)$ and output $y(0), \ldots y(k-1)$ measurements. This allows the definition of the state sequence that is recovered by 4SID methods with $\hat{X}_0$ as initial states:

$$\hat{X}_f = \left(\hat{x}(i)\, \hat{x}(i+1) \ldots \hat{x}(i+j-1) \right)$$

$$= \left(A^k - \Omega_i \Gamma_i \mid \Delta_i^d - \Omega_i H_i^d \mid \Omega_i \right) \begin{pmatrix} \hat{X}_0 \\ \hline U_p \\ \hline Y_p \end{pmatrix} \tag{10.52}$$

$$= \left(A^k - \Omega_i \Gamma_i \mid \Delta_i^d - \Omega_i H_i^d \quad \Omega_i \right) \begin{pmatrix} \hat{X}_0 \\ \hline W_p \end{pmatrix}$$

This state sequence is generated by a bank of non-steady-state Kalman filters, working in parallel on each of the columns of the block Hankel matrix of past inputs and outputs W_p. As can be seen, each state is filtered only from the limited length of past input/output information (i samples), which is contained in one column of W_p. Therefore the initial state estimate:

$$\hat{X}_0 = \left(\hat{x}^0(i)\, \hat{x}^0(i+1) \ldots \hat{x}^0(j-1) \right) \tag{10.53}$$

plays important role and advanced 4SID methods are shown to use nonzero initial estimates. For example the described N4SID method uses:

$$\hat{X}_0 = X_p^d \underset{U_f}{/} U_p, \tag{10.54}$$

where X_p^d is the deterministic component of the state sequence matrix X_p. One has to keep in mind that this is only an interpretation of 4SID methods from the Kalman filter perspective. For example the prior estimate $\hat{X}_0 = X_p^d /_{U_f} U_p$ is never computed in the 4SID algorithm, but it can be shown that a bank of nonsteady-state KF with this prior estimate and working with the input/output data from W_p columns will give the same state sequence estimate as the N4SID algorithm. The prior estimate $\hat{X}_0 = X_p^d /_{U_f} U_p$ cannot be even explicitly computed, because for example X_p^d is a priori unknown. The major observation in the subspace algorithms is that the system matrices A, B, C, D, Q, R, and S do not have to be known to determine the state sequence. It can be determined directly from the input/output data.

References

Ho, B.L., Kalman, R.E., 1966. Editorial: effective construction of linear state-variable models from input/output functions. Automatisierungstechnik 14, 545–548.

Van Overschee, P., De Moor, B., 1996. Subspace Identification for Linear Systems: Theory-Implementation-Applications. Kluwer Academic Publishers.

Chapter 11

New applications in chemometrics

Chapter outline

11.1 Limitations of the Kalman filter and new directions

As discussed in the previous chapters, when a system and its observations are represented in state space with linear equations, with system noise and measurement noise white, Gaussian, and mutually uncorrelated, and the system and measurement noise statistics are exactly known, a Kalman filter (KF) with the same order as the system provides optimal state estimates in a way that is simple and fast with minimal computation and memory requirements. As discussed in Kwon and Han (2005), however, the KF is an infinite impulse response (IIR) filter. This implies that the KF performance may be poor under certain operational conditions (Jazwinski, 1970). For example, it is more susceptible to problems of finite-length arithmetic, such as noise generated by calculations, and limit cycles. This is a direct consequence of feedback: when the output is not computed perfectly and is fed back, the imperfections can compound. Furthermore, there are associated difficulties to implement IIR filters using fixed-point arithmetic (slower), and they lack the computational advantages of finite impulse response (FIR) filters for multirate (decimation and interpolation) applications. Other issues often arising are due to insufficient robustness against mismodeling of the underlying process (Hanlon and Maybeck, 2000), especially when the studied processes cannot be accurately approximated with a reasonable number of states. Other issues are associated with temporal uncertainties (Kwon and Han, 2005), a strong dependence on the filter performance in the initial values and a high vulnerability to errors in the noise statistics (Gibbs, 2011;

State Estimation in Chemometrics. https://doi.org/10.1016/B978-0-08-102603-8.00011-0

Daum, 2005; Simon, 2006; Shmaliy, 2009). These issues are further elaborated as follows.

An issue arising from the mismodeling of a chemical process is that it may cause an estimator to be biased and/or diverge (Simon, 2006). Even if a process is represented in state space accurately on a long time scale, it may undergo unpredictable changes, such as jumps in frequency, phase, and velocity. Because these effects typically manifest themselves within a short time horizon, they are called temporal uncertainties. To diminish the effects of mismodeling and temporal uncertainties, an estimator should therefore be robust. However, practice shows that the KF is less robust than batch estimators (Jazwinski, 1968; Kwon and Han, 2005; Shmaliy, 2011). This fact is further explained in the work by Jazwinski (1968), which discusses the finite impulse response (FIR) of a batch estimator, such structure does not allow projecting "old" errors beyond the averaging horizon to the estimate. In spite of the numerous modifications that have taken place over the years, the KF remains a recursive estimator and is of infinite impulse response (IIR). The problems that most often arise when attempting to use the KF in the real world are due to its IIR structure (Jazwinski, 1968). It seems that a limited memory filter appears to be the only device for preventing divergence in the presence of unbounded perturbations in the system (Jazwinski, 1970).

Another issue that is also worth noting is that almost all of the early approaches to KF have been relying on limited data and were developed to increase the robustness of the KF and EKF to mismodeling, uncertain conditions, and poor information about noise. However, the filters of this kind have an FIR structure, for which the straightforward form is convolution based. Employing the convolution, receding-horizon (RH) FIR filtering has been developed in Kwon and Han (2005), both in the batch and iterative forms, to provide one-step predictive estimates required in model predictive control (Kucera, 1991). Most RH FIR algorithms were designed to be bias constrained (Kwon and Han, 2005; Kwon et al., 1989). Among them, the minimum-variance FIR filter subject to the unbiasedness constraint (Kwon et al., 2002) has demonstrated the widest adoption. Further developments of RH FIR filtering can be found in the works by Han et al. (2002), Ahn (2012, 2014), Kim (2010), Kim and Lee (2007), and Ahn et al. (2006, 2016). Again, these advances may be effectively adopted for use in a chemometrics context and process control applications.

As mentioned earlier, in order for the KF to be optimal, the noise statistics must be known. However, noise measurements are costly and hard to conduct for chemical time-varying systems (Gibbs, 2011). The noise statistics can be defined through approximations (Rong and Bar-Shalom, 1994; Sarkka and Nummenmaa, 2009) using some prior knowledge about the chemical process. However, if the approximation errors are too large, the KF estimate value may be too far from optimal (Jazwinski, 1970). At this point, it is also worth noting that noise in some applications is nonwhite and non-Gaussian and may even have a heavy tail, as in many industrial process control applications (Auger et al., 2013), which may also severely affect the KF estimates. Because of this, many modifications to the original Kalman filter have therefore been proposed over the past few decades to improve performance. The most significant modifications are the extended Kalman filter (EKF) derived for nonlinear systems (Cox, 1964) and the unscented Kalman filter (UKF) proposed for highly nonlinear systems (Julier and Uhlmann, 1997). Among other important modifications are the fading-memory version of the KF (which has improved robustness properties (Sorenson and Sacks, 1971)), the invariant extended Kalman filter (Bonnabel et al., 2009) for nonlinear systems possessing symmetries (or invariances), the ensemble Kalman filter (Evensen, 1994) for high-dimensional systems, the fast Kalman filter (Lange, 2001) for systems with sparse matrices, and the robust Kalman filter (Masreliez, 1975; Martin and Masreliez, 1975; Maronna and Yohai, 2006) for models with heavy-tailed noise or Gaussian noise mixed with outliers. Some further modifications include algorithms for uncertain observations (Nahi, 1969), algorithms for multiresolution stochastic processes (Basseville et al., 1992; Baccarelli and Cusani, 1996; Ait-El-Fquih and Desbouvries, 2006), algorithms for compressed sensing (Vaswani, 2008), and embedded pseudonorms (Carmi et al., 2010). In addition to these, there are also algorithms that can be used in the chemical industry, which utilize the signals associated with bounded disturbances (Becis-Aubry et al., 2008; Rawicz et al., 2003) or smoothing filtering techniques (Simon and Shmaliy, 2013) or distributed filtering techniques that make use of consensus strategies (Carli et al., 2008) or algorithms for systems with bilinear dynamics (Luenberger, 1964; Gauthier et al., 1992a,b), which may also be adapted for use in a chemometrics context. A comprehensive list of KF algorithms can be found in the work by Simon (2006).

11.2 Recent advances with the fractional-order Kalman filter

Fractional calculus is dealing with integration and differentiation of noninteger order (Odlham and Spaniar, 1974; Podlubny, 1999; Hilfer, 2000; Monje et al., 2010; Abdeljawad, 2011; Machado et al., 2014; Dadras et al., 2017). Tustin discretization methods were discussed in the work by Vinagre et al. (2003). The stability of fractional-order systems has also been extensively investigated (Caponetto and Dongola, 2013). The method has been widely used as a tool to model processes in industrial automation (Efe, 2011; Sierociuk et al., 2011), neural networks (Wang et al., 2014), and diffusion wave analysis (El-Sayed, 1996). Another application of fractional-order calculus, rapidly developing nowadays in process control, is the design of fractional-order proportional integral derivative (PID) controllers with general structure, $PI^\lambda D^\mu$ where parameters λ, μ are of noninteger value (Podlubny, 1999). The fractional-order PID is proved to be a much more flexible approach in the controller design, with respect to the standard PID, as there are only five parameters that need to be tuned instead of three, as discussed in Cao and Cao (2006), Hamamci (2007), Bettayeb and Mansouri (2014), and references therein. More importantly, such structure enables process engineers to design much simpler controllers instead of using high-order designs that would be required to control processes with more complex high-order dynamics. In addition, fractional-order sliding mode controllers (Yin et al., 2014, 2015, 2017) and fractional-order iterative learning controllers (Li et al., 2011) may be designed for more complex chemical processes.

As the number of applications has increased over the past several years, discrete fractional Kalman filter formulations have already been proposed (Najar et al., 2009; Torabi et al., 2016). Furthermore, Simon (2010) proposed a formulation of Kalman filtering with state constraints. In the work by Wang et al. (2013), a numerical method for nonlinear fractional-order differential equations with constant or time-varying delay was proposed. In addition, Yang and Cao (2013) discussed the existence and uniqueness of initial value problems for nonlinear higher-order fractional equations with delay, these were studied by fixed point theory. Using the Lyapunov-Krasovskii functional, Rakkiyappan et al. (2014) adopted the delay-fractional approach and the stochastically asymptotic stability of a neural network was considered by solving a set of linear matrix inequalities. The robust stability for fractional-order linear time-invariant system was investigated in Liao et al. (2011), the approach adopted a deterministic linear coupling relationship between fractional-order and other model parameters. In another study, Wei et al. (2009) considered the robust filtering problem for a class of discrete-time uncertain stochastic nonlinear time-delay systems with both the probabilistic missing measurements and external stochastic disturbances. In the work by Ahmad and Namerikawa (2013), the extended Kalman filter with intermittent measurements was considered.

In Hu et al. (2013), the authors considered the recursive finite-horizon filtering problem for a class of nonlinear time-varying systems subject to multiplicative noises, missing measurements, and quantization effects. Compared with integer-order Kalman filters (Kalman, 1960; Brown et al., 1997; Shen et al., 2008; Zhang et al., 2014; Liang et al., 2014), the fractional ones show better performances to satisfy the requirements of the designers, a detailed introduction was provided in the work by Sierociuk and Dzieliński (2006). By using the cumulative vector and extended matrix approach, Sadeghian and Salarieh (2013) developed an alternative method to derive the fractional-order Kalman filter (FKF). The fractional-order Kalman filter can also be used in singular systems and systems containing subsystems (Ashayeri et al., 2013). In the work by Sun et al. (2013b), a Kalman smoother was combined with a weighted fusion algorithm to address problems associated with weighted measurement fusion smoothing problems for fractional systems. Furthermore, the fractional-order extended Kalman filter (EKF) and the fractional-order unscented Kalman filter (UKF) were similarly used to deal with nonlinear systems (Águila et al., 2012). A detailed derivation for the EKF was discussed in the work by Sierociuk and Dzieliński (2006). The unscented fractional-order Kalman filter is also an effective way to realize the robust state estimation for nonlinear fractional-order systems, and the corresponding algorithm was provided in the paper by Sierociuk et al. (2016).

At this point, it is worth noting that non-Gaussian noises, especially Lévy noises, are widely encountered in chemical systems and process control applications. Theoretically, Lévy noises, however, can be approximated by the increments of the corresponding Lévy process per time step (Applebaum, 2009). More recently, for the non-Gaussian noise case, fractional-order Kalman filters for Lévy noise and colored noise were discussed in the work by Sierociuk and Ziubinski (2014) as well as the paper by Wu et al. (2015), respectively. An issue that often arises in this case is that due to the infinite variance, it is difficult to estimate the system states by using traditional algorithms when considering Lévy noises. Sun et al. (2013a,b) investigated the KF algorithm with Lévy noises instead of Gaussian white noises, but their analysis did not involve the fractional-order systems. Furthermore, both FKF and EKF (or UKF) results are all based on the assumption that the system noise and measurement noise are all the Gaussian white noises. Finally the case of FKF with non-Gaussian noises has been addressed in Wu et al. (2015). The extended Kalman filter for a nonlinear fractional-order system with colored measurement noise was proposed in the work by Safarinejadian et al. (2018) and more recently by Gao (2018) where he proposed Tustin generating functions to approximate the dynamics of a continuous-time fractional-order system. Correlations of process and measurement noises using fractional-order Kalman filters were investigated in the works by Pourdehi et al. (2015) and Sierociuk (2013).

For the formulation of the fractional-order Kalman filter, one needs to start with the fractional-order Grünwald-Letnikov difference:

$$\Delta^{\alpha}x(k) = \frac{1}{h^{\alpha}}\sum_{j=0}^{k}(-1)^{j}\binom{\alpha}{j}x(k-j) \qquad (11.1)$$

where Δ is the fractional system-order operator, α is the fractional order, h is the sampling time, and k is the number of samples (also step size). Furthermore:

$$\binom{\alpha}{j} = \begin{cases} 1 & \text{if } j = 0 \\ \dfrac{\alpha(\alpha-1)...(\alpha-j+1)}{j!} & \text{if } j > 0 \end{cases} \qquad (11.2)$$

The general discrete linear fractional-order system can be formulated as follows:

$$\Delta^{\alpha}x(k+1) = Ax(k) + Bu(k) + w(k) \qquad (11.3)$$

$$x(k+1) = \Delta^\alpha x(k+1) - \sum_{j=0}^{k+1} (-1)^j \gamma_j x(k+1-j) \tag{11.4}$$

$$y(k) = Cx(k) + \nu(k) \tag{11.5}$$

where $x(k)$ is the state vector, $u(k)$ is the system input, $y(k)$ is the measurement output, A, B, and C are the known constant matrices with appropriate dimensions, $w(k)$ and $v(k)$ represent system noise and measurement noise at time instant k, respectively, which are always assumed to be the Gaussian white noises with zero means characterized by the following covariance matrices:

$$E\left[w_i w_j^T\right] = \begin{cases} Q_i & \text{if } i=j \\ 0 & \text{if } i \neq j \end{cases} \tag{11.6}$$

$$E\left[\nu_i \nu_j^T\right] = \begin{cases} R_i & \text{if } i=j \\ 0 & \text{if } i \neq j \end{cases} \tag{11.7}$$

$$\gamma_j = diag\left(\left[\binom{\alpha_1}{j}\binom{\alpha_2}{j}\cdots\binom{\alpha_N}{j}\right]\right) \tag{11.8}$$

where Q_i is the system noise covariance matrix at time instant i, R_i is the measurement noise covariance matrix at time instant i: α_1, α_2, ..., α_N are the orders of the fractional-order system and $j = 1, 2, ..., k+1$.

The state prediction vector at time instant k is:

$$\tilde{x}(k) = E[x(k)|y^*(k-1)] \tag{11.9}$$

where $y^*(k-1)$ is the measurement sequence containing values of the measurement output $y(0)$, $y(1)$, ..., $y(k-1)$ and system input signal $u(0)$, $u(1)$, ..., $u(k-1)$. The state estimation vector at time instant k is:

$$\hat{x}(k) = E[x(k)|y^*(k)] \tag{11.10}$$

and the corresponding prediction error covariance matrix and estimation error covariance matrix at time instant k are given by:

$$\tilde{p}(k) = \begin{cases} E\left[(x(i)-\tilde{x}(i))(x(k)-\tilde{x}(k))^T\right] & \text{if } i=k \\ 0 & \text{if } i \neq k \end{cases}$$

$$\hat{p}(k) = \begin{cases} E\left[(x(i)-\hat{x}(i))(x(k)-\hat{x}(k))^T\right] & \text{if } i=k \\ 0 & \text{if } i \neq k \end{cases} \tag{11.11}$$

Following the earlier content, for the discrete linear fractional-order system, the fractional-order Kalman filter is given by the following sets of formulas:

$$\Delta^\alpha \tilde{x}(k+1) = A\hat{x}(k) + Bu(k) \tag{11.12}$$

$$\tilde{x}(k+1) = \Delta^\alpha \tilde{x}(k+1) - \sum_{j=1}^{k+1} (-1)^j \gamma_j \hat{x}(k+1-j) \qquad (11.13)$$

$$\tilde{p}(k+1) = (A+\gamma_1)\hat{p}(k)(A+\gamma_1)^T + Q(k) + \sum_{j=2}^{k+1} \gamma_j \hat{p}(k+1-j)\gamma_j^T \qquad (11.14)$$

The algorithm is summarized by the following flowchart in Fig. 11.1.

$$\hat{x}(k) = \tilde{x}(k) + K(k)(y(k) - C\tilde{x}(k)) \qquad (11.15)$$

$$\hat{p}(k) = (I - K(k)C)\tilde{p}(k) \qquad (11.16)$$

where:

$$K(k) = \tilde{p}(k)C^T \left(C\tilde{p}(k)C^T + R(k) \right)^{-1} \qquad (11.17)$$

with initial conditions $\tilde{x}(0)$, $\tilde{p}(0)$, $\nu(k)$, $w(k)$ assumed to be independent and with zero expected value.

Through the analysis provided so far, based on the obtained FKF for a fractional-order system, some problems, such as state estimation, parameter identification, etc., could be realized with a satisfactory accuracy. However, in this FKF algorithm, the system noise and the measurement noise are assumed to be Gaussian and white, and this assumption is not always realistic in practice. Unfortunately, the obtained FKF cannot obtain a satisfactory performance if the system and measurement noises are not Gaussian white noises; thus, it is necessary to often use a modified fractional-order Kalman filter for fractional-order system that have Lévy noises. At this point, it is worth reminding the readers that a Lévy process consists of the sum of a Brownian motion process and a pure jump process. But the Gaussian process can be approximated by the increments of Brownian motion per time step. This observation enables us to decompose a Lévy noise into a Gaussian white noise with some extremely large values.

Let $\bar{w}(k)$ and $\bar{\nu}(k)$ be the system Lévy noise and the measurement Lévy noise with extremely large values, respectively, at time instant k. The discrete linear fractional-order system with Lévy noises can be formulated as follows:

$$\Delta^\alpha x(k+1) = Ax(k) + Bu(k) + \bar{w}(k) \qquad (11.18)$$

$$x(k+1) = \Delta^\alpha x(k+1) - \sum_{j=0}^{k+1} (-1)^j \gamma_j x(k+1-j) \qquad (11.19)$$

$$y(k) = Cx(k) + \bar{\nu}(k) \qquad (11.20)$$

Due to the fact that noises $\bar{w}(k)$ and $\bar{\nu}(k)$ are Lévy noises, it is difficult to obtain $\bar{x}(k+1)$ and $\bar{y}(k)$ directly. Using the fact that the Lévy noise can be

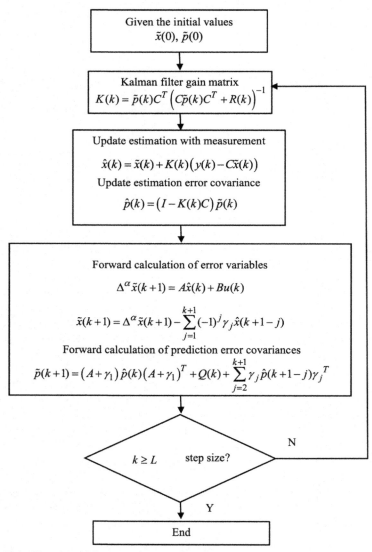

FIG. 11.1   Flowchart of the fractional-order Kalman filter algorithm (Wu et al., 2015).

decomposed into a Gaussian white noise with some extremely large values, similarly, $\bar{x}(k+1)$ and $\bar{y}(k)$ could be approximated by the following expressions:

$$\bar{x}(k+1) = \begin{cases} \Omega_1 + \delta_1 \cdot \text{sign}(\Omega_2) & \text{if } |\Omega_2| \geq \delta_1 \\ x(k+1) & \text{if } |\Omega_2| < \delta_1 \end{cases} \quad (11.21)$$

$$\bar{y}(k) = \begin{cases} C\tilde{x}(k) + \delta_2 \cdot \text{sign}(y(k) - C\tilde{x}(k)) & \text{if } |y(k) - C\tilde{x}(k)| \geq \delta_2 \\ y(k) & \text{if } |y(k) - C\tilde{x}(k)| < \delta_2 \end{cases} \quad (11.22)$$

where:

$$\Omega_1 = A\tilde{x}(k) + Bu(k) - \sum_{j=0}^{k+1} (-1)^j \gamma_j \tilde{x}(k+1-j) \quad (11.23)$$

$$\Omega_2 = x(k+1) - A\tilde{x}(k) - Bu(k) + \sum_{j=1}^{k+1} (-1)^j \gamma_j \tilde{x}(k+1-j) \quad (11.24)$$

and the $\tilde{x}(k)$ in Ω_1 is the state prediction vector at time instant k, and the $x(k+1)$ in Ω_2 is the state vector at time instant $k+1$ under Lévy noises, δ_1 and δ_2 are positive threshold values, respectively.

If the absolute value of the approximated system noise Ω_2 is less than the given threshold value δ_1, the approximated state value $\bar{x}(k+1)$ at time instant $k+1$ is equal to the real state value under non-Gaussian Lévy noise; otherwise, it would be evaluated by Ω_1 plus or minus the threshold value δ_1. It is assumed to be universally approximating the measurement value $\bar{y}(k)$. If the absolute value of the approximated measurement noise $(y(k) - C\tilde{x}(k))$ is less than the given threshold value δ_2, the approximated measurement value $\bar{y}(k)$ at time instant k is equal to the real measurement value; otherwise, it is equal to $C\tilde{x}(k)$ plus or minus the threshold value δ_2. The piecewise equations of $\bar{x}(k+1)$ and $\bar{y}(k)$ enable us to obtain the corresponding values based on previous predicted values. The expressions for the Kalman filter algorithm are accordingly modified as follows:

$$\Delta^\alpha \tilde{x}(k+1) = A\hat{x}(k) + Bu(k) \quad (11.25)$$

$$\tilde{x}(k+1) = \Delta^\alpha \tilde{x}(k+1) - \sum_{j=1}^{k+1} (-1)^j \gamma_j \hat{x}(k+1-j) \quad (11.26)$$

$$\tilde{p}(k+1) = (A+\gamma_1)\hat{p}(k)(A+\gamma_1)^T + \overline{Q}(k) + \sum_{j=2}^{k+1} \gamma_j \hat{p}(k+1-j)\gamma_j^T \quad (11.27)$$

$$\hat{x}(k) = \tilde{x}(k) + K(k)(\bar{y}(k) - C\tilde{x}(k)) \quad (11.28)$$

$$\hat{p}(k) = (I - K(k)C)\tilde{p}(k) \quad (11.29)$$

where:

$$\overline{Q}(k) = \left(\bar{x}(k+1) - A\tilde{x}(k) - Bu(k) + \sum_{j=1}^{k+1}(-1)^j \gamma_j \tilde{x}(k+1-j) \right)$$

$$\times \left(\bar{x}(k+1) - A\tilde{x}(k) - Bu(k) + \sum_{j=1}^{k+1}(-1)^j \gamma_j \tilde{x}(k+1-j) \right)^T \qquad (11.30)$$

$$+ (A+\gamma_1)\tilde{p}(k)(A+\gamma_1)^T + \sum_{j=1}^{k+1} \gamma_j \tilde{p}(k+1-j)\,\gamma_j^T$$

and:

$$\overline{R}(k) = (\bar{y}(k) - C\tilde{x}(k))(\bar{y}(k) - C\tilde{x}(k))^T + C\tilde{p}(k)C^T \qquad (11.31)$$

The algorithm is summarized by the following flowchart in Fig. 11.2.

11.3   Kalman filter observability perspective

Formally, a process is said to be observable if for any possible control vectors and state sequence, the current state can be determined in finite time using only the outputs. For a process that is not observable, the current values of some of its states cannot be determined through output sensors. This implies that the state values will also be unknown to the controller so the latter will be unable to fulfill its specifications. In mathematical terms, the nonlinear system may be described in state space as follows:

$$\begin{aligned} x(t) &= f(x(t),\ u(t)) \\ y(t) &= h(x(t)) \end{aligned} \qquad (11.32)$$

where functions f and h are continuous and differentiable and the input functions are locally bounded. As discussed earlier, in the case of linear systems, if the observability matrix is invertible and full rank, the state vector is observable from the output (Luenberger, 1971). For nonlinear systems, equivalent observability criteria can be invoked (Keller, 1987; Diop and Filess, 1991; Gauthier et al., 1992a,b; Bogaerts, 1999; Gouze et al., 2000; Diop, 2001; Boizot et al., 2010; Chyi-Tsong and Shih-Tien, 2005; Cruz-Victoria et al., 2008).

One can rewrite the system equations following linearization using Jacobian matrices (Fliess, 1990, 1995; Meta-Machuca et al., 1990; Isidori, 1999; Guan et al., 2001; Khalil, 2002; Martínez-Guerra and Diop, 2004). If these matrices are defined as $J(x, u)$, $H(x)$:

$$\begin{aligned} \dot{x}_L(t) &= J(x(t),\ u(t))x_L(t) \\ y_L(t) &= H(x(t))x_L(t) \end{aligned} \qquad (11.33)$$

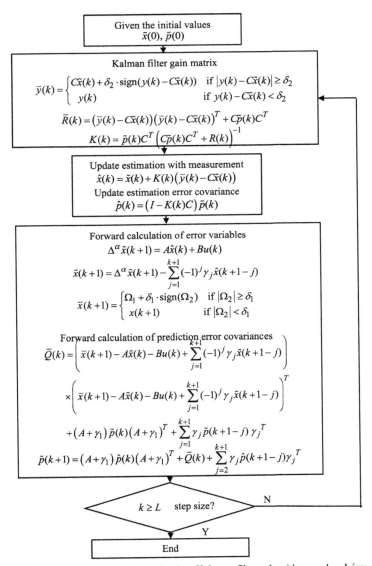

FIG. 11.2 Flowchart of the fractional-order Kalman filter algorithm under Lévy noises (Wu et al., 2015).

A similar observability matrix $\Lambda = [H, HJ, HJ^2, \ldots HJ^n]^T, \det \Lambda \neq 0$ for the linearized system is generated, this must be nonsingular in order to assure full observability conditions so as to obtain $x = \Lambda^{-1}y$. The linearization is based on diffeomorphic transformation employing Lie derivatives of the nonlinear system into an equivalent linear system through a change of variables and a suitable control input. This transformation can be applied to nonlinear systems of the form:

$$\dot{x} = f(x) + g(x)u$$
$$y = h(x), \ \dot{y} = L_f h(x), \ddot{y} = L_f^2 h(x), \ldots, y^{n-1} = L_f^{n-1} h(x) \tag{11.34}$$
$$y^n = L_f^n h(x) + L_g L_f^{n-1} h(x)u$$

where L_f^r is the rth order Lie derivatives, which are the directional derivatives of the corresponding state variables along the measured output trajectory. $dL_f^r h$ are the differentials of the rth order Lie derivative recursively defined as follows:

$$L_f^0 h \doteq h, \ dL_f^0 h \doteq dh = \left(\frac{\partial h}{\partial x_1}, \ \cdots, \ \frac{\partial h}{\partial x_n} \right)$$

$$L_f^1 h \doteq \langle dh, f \rangle = \sum_{i=1}^n \frac{\partial h}{\partial x_1} f_i, \ dL_f^1 h \doteq \left(\frac{\partial}{\partial x_1} \left(\sum_{i=1}^n \frac{\partial h}{\partial x_1} f_i \right), \ \cdots, \ \frac{\partial}{\partial x_n} \left(\sum_{i=1}^n \frac{\partial h}{\partial x_1} f_i \right) \right)$$

$$L_f^r h \doteq \left\langle dL_f^{r-1} h, \ f \right\rangle = L_f \left(L_f^{r-1} h \right), \ r \geq 2$$

$$\tag{11.35}$$

The coordinate transformation $T(x)$ that carried out the system into normal form comes from the first $(n-1)$ derivatives:

$$T(x) = \begin{pmatrix} y = h(x) \\ \dot{y} = L_f h(x) \\ \vdots \\ {}^{n-1}y = L_f^{n-1} h(x) \end{pmatrix} \tag{11.36}$$

and transforms trajectories from the original x coordinate system into the new coordinate system. As long as the transformation is a diffeomorphism, smooth trajectories in the original coordinate system will have unique counterparts in the new coordinate system that are also smooth.

In order to clarify the procedure, one can consider an example of a nonisothermal nonadiabatic stirred chemical reactor. Such systems are addressed often in the bioreactors literature, e.g., Bastin and Dochain (1990), Betancur et al. (2006). Assuming a second-order chemical reaction $A + B \to P$ takes place so that:

$$\dot{x}_1 = f_1(x_1, x_2, x_3) = \alpha(a - x_1) - x_1 x_2 \exp\left(-\frac{\beta}{x_3} \right)$$

$$\dot{x}_2 = f_2(x_1, x_2, x_3) = \alpha(\delta - x_2) - x_1 x_2 \exp\left(-\frac{\beta}{x_3} \right) \tag{11.37}$$

$$\dot{x}_3 = f_3(x_1, x_2, x_3) = b(\rho - x_3) - d x_1 x_2 \exp\left(-\frac{\beta}{x_3} \right) - \gamma(u - x_3)$$

$$y = x_3$$

where x_1 is the mass balance of reactive species A, x_2 denotes the mass balance of reactive species B, x_3 represents the energy balance, y is the measured output, u is a constant input, α is the inlet concentration reactive A, β is the activation energy, δ is the inlet concentration reactive B, and ρ is the inlet enthalpy.

For a linearized version of the system:

$$\dot{x} = Ax,$$

$$x^T = (x_1, x_2, x_3)$$

$$A = \begin{pmatrix} \dfrac{\partial f_1}{\partial x_1} & \dfrac{\partial f_1}{\partial x_2} & \dfrac{\partial f_1}{\partial x_3} \\[2mm] \dfrac{\partial f_2}{\partial x_1} & \dfrac{\partial f_2}{\partial x_2} & \dfrac{\partial f_2}{\partial x_3} \\[2mm] \dfrac{\partial f_3}{\partial x_1} & \dfrac{\partial f_3}{\partial x_2} & \dfrac{\partial f_3}{\partial x_3} \end{pmatrix} \tag{11.38}$$

with:

$$y = x_3$$

$$\frac{\partial f_1}{\partial x_1} = -a - x_2 \exp\left(-\frac{\beta}{x_3}\right) \quad \frac{\partial f_1}{\partial x_2} = -x_1 \exp\left(-\frac{\beta}{x_3}\right) \quad \frac{\partial f_1}{\partial x_3} = \beta\frac{x_1 x_2}{x_3^2} \exp\left(-\frac{\beta}{x_3}\right)$$

$$\frac{\partial f_2}{\partial x_1} = -x_2 \exp\left(-\frac{\beta}{x_3}\right) \quad \frac{\partial f_2}{\partial x_2} = -a - x_1 \exp\left(-\frac{\beta}{x_3}\right) \quad \frac{\partial f_2}{\partial x_3} = \beta\frac{x_1 x_2}{x_3^2} \exp\left(-\frac{\beta}{x_3}\right)$$

$$\frac{\partial f_3}{\partial x_1} = -dx_2 \exp\left(-\frac{\beta}{x_3}\right) \quad \frac{\partial f_3}{\partial x_2} = -dx_1 \exp\left(-\frac{\beta}{x_3}\right) \quad \frac{\partial f_3}{\partial x_3} = -b + d\beta\frac{x_1 x_2}{x_3^2} \exp\left(-\frac{\beta}{x_3}\right) + \gamma$$

$$\tag{11.39}$$

where the corresponding observability matrix is given by:

$$\begin{pmatrix} C \\ CA \\ CA^2 \end{pmatrix} = \begin{pmatrix} 0 & \dfrac{\partial f_3}{\partial x_1} & \dfrac{\partial f_1}{\partial x_1}\dfrac{\partial f_3}{\partial x_1} + \dfrac{\partial f_2}{\partial x_1}\dfrac{\partial f_3}{\partial x_2} + \dfrac{\partial f_3}{\partial x_1}\dfrac{\partial f_3}{\partial x_3} \\[3mm] 0 & \dfrac{\partial f_3}{\partial x_2} & \dfrac{\partial f_1}{\partial x_2}\dfrac{\partial f_3}{\partial x_1} + \dfrac{\partial f_2}{\partial x_2}\dfrac{\partial f_3}{\partial x_2} + \dfrac{\partial f_3}{\partial x_2}\dfrac{\partial f_3}{\partial x_3} \\[3mm] 1 & \dfrac{\partial f_3}{\partial x_3} & \dfrac{\partial f_1}{\partial x_3}\dfrac{\partial f_3}{\partial x_1} + \dfrac{\partial f_2}{\partial x_3}\dfrac{\partial f_3}{\partial x_2} + \left(\dfrac{\partial f_3}{\partial x_2}\right)^2 \end{pmatrix} \tag{11.40}$$

where the determinant of the observability matrix is different from zero:

$$\left(\frac{\partial f_1}{\partial x_2}\frac{\partial f_3}{\partial x_1} + \frac{\partial f_2}{\partial x_2}\frac{\partial f_3}{\partial x_2} + \frac{\partial f_3}{\partial x_2}\frac{\partial f_3}{\partial x_3}\right)\frac{\partial f_3}{\partial x_1} - \left(\frac{\partial f_1}{\partial x_1}\frac{\partial f_3}{\partial x_1} + \frac{\partial f_2}{\partial x_1}\frac{\partial f_3}{\partial x_2} + \frac{\partial f_3}{\partial x_1}\frac{\partial f_3}{\partial x_3}\right)\frac{\partial f_3}{\partial x_2} \neq 0$$

From a differential-geometric observability perspective, the transformed system can be written as:

$$\vartheta = \left(dL_j^0 h \ dL_j^1 h \ dL_j^2 h \right)^T \tag{11.41}$$

and for an output $y = x_3$:

$$\Phi_{1r} = dL_f^0 h = x_3 \tag{11.42}$$

$$\Phi_{2r} = dL_f^1 h = b(\rho - x_3) - dx_1 x_2 \exp\left(-\frac{\beta}{x_3}\right) - \gamma(u - x_3) \tag{11.43}$$

$$\Phi_{3r} = dL_f^2 h = \omega \zeta \tag{11.44}$$

where:

$$\omega = (\omega_1 \quad \omega_2 \quad \omega_3)$$
$$= \left(-dx_1 \exp\left(-\frac{\beta}{x_3}\right) \quad -dx_2 \exp\left(-\frac{\beta}{x_3}\right) \quad -b + d\beta\frac{x_1}{x_3^2}\exp\left(-\frac{\beta}{x_3}\right) + \gamma\right) \tag{11.45}$$

$$\zeta = \begin{pmatrix} \zeta_1 \\ \zeta_2 \\ \zeta_3 \end{pmatrix} = \begin{pmatrix} \alpha(a - x_1) - x_1 x_2 \exp\left(-\dfrac{\beta}{x_3}\right) \\[2mm] \alpha(\delta - x_2) - x_1 x_2 \exp\left(-\dfrac{\beta}{x_3}\right) \\[2mm] b(\rho - x_3) - dx_1 x_2 \exp\left(-\dfrac{\beta}{x_3}\right) - \gamma(u - x_3) \end{pmatrix} \tag{11.46}$$

so that:

$$\vartheta = \left(\frac{\partial \Phi_r}{\partial x}\right) = \begin{pmatrix} \dfrac{\partial \Phi_{1r}}{\partial x_1} & \dfrac{\partial \Phi_{1r1}}{\partial x_2} & \dfrac{\partial \Phi_{1r1}}{\partial x_3} \\[2mm] \dfrac{\partial \Phi_{2r}}{\partial x_1} & \dfrac{\partial \Phi_{2r}}{\partial x_2} & \dfrac{\partial \Phi_{2r}}{\partial x_3} \\[2mm] \dfrac{\partial \Phi_{3r}}{\partial x_1} & \dfrac{\partial \Phi_{3r}}{\partial x_2} & \dfrac{\partial \Phi_{3r}}{\partial x_3} \end{pmatrix} = \begin{pmatrix} 0 & 0 & 1 \\[2mm] \dfrac{\partial \Phi_{2r}}{\partial x_1} & \dfrac{\partial \Phi_{2r}}{\partial x_2} & \dfrac{\partial \Phi_{2r}}{\partial x_3} \\[2mm] \dfrac{\partial \Phi_{3r}}{\partial x_1} & \dfrac{\partial \Phi_{3r}}{\partial x_2} & \dfrac{\partial \Phi_{3r}}{\partial x_3} \end{pmatrix} \tag{11.47}$$

where:

$$\frac{\partial \Phi_{2r}}{\partial x_1} = -dx_2 \exp\left(-\frac{\beta}{x_3}\right)$$

$$\frac{\partial \Phi_{2r}}{\partial x_2} = -dx_1 \exp\left(-\frac{\beta}{x_3}\right) \tag{11.48}$$

$$\frac{\partial \Phi_{2r}}{\partial x_3} = -b + d\beta\frac{x_1 x_2}{x_3^2}\exp\left(-\frac{\beta}{x_3}\right) + \gamma$$

$$\frac{\partial \Phi_{3r}}{\partial x_j} = -\sum_{i=1}^{3}\left(\frac{\partial \omega_i}{\partial x_j}\zeta_i + \omega_i\frac{\partial \zeta_i}{\partial x_j}\right), \quad j = 1, 2, 3 \tag{11.49}$$

For the system to be observable:

$$\text{rank}\left(\frac{\partial \Phi_r}{\partial x}\right) = 3 \tag{11.50}$$

Formulations of the extended Kalman filter using Lie groups can be found in the comprehensive work by Bourmaud et al. (2016).

11.4 System analysis based on sum of squares techniques

Beyond the Kalman filter, the sum of squares (SOS) decomposition (Parrilo, 2000; Parrilo and Lall, 2003; Parrilo and Sturmfels, 2001) is a particularly useful generic tool for the analysis of chemical systems described by nonlinear ordinary differential equations or differential algebraic equations, for hybrid chemical systems with nonlinear subsystems and/or nonlinear switching processes as encountered in chemical engineering and systems biology, and for time-delay systems described by nonlinear functional differential equations.

The observation that the SOS decomposition can be computed efficiently using semidefinite programming has introduced the need for developing software that would facilitate the formulation of the semidefinite programs from their SOS equivalents. One such software is SOSTOOLS (Papachristodoulou et al., 2013) a free, third-party MATLAB toolbox for solving SOS programs. SOS solves a convex optimization problem of the form:

$$\text{Minimize} \sum_{j=1}^{J} w_j c_j \text{ subject to} : a_{i,0}(x) + \sum_{j=1}^{J} a_{i,j}(x)c_j, \quad i = 1,...I \tag{11.51}$$

where the c_j's are the scalar real decision variables, the w_j's are real numbers, and the $a_{i,j}(x)$'s are some given polynomials with fixed coefficients. While the conversion from SOS programs to semidefinite programs (SDPs) can be manually performed for small-size instances or tailored for specific problem classes, such a conversion can be quite difficult in general (Vandenberghe and Boyd, 1996). It is therefore desirable to have a computational aid that automatically performs this conversion for general SOS programs. SOSTOOLS automates the conversion from SOS program to SDP, calls the SDP solver, and converts the SDP solution back to the solution of the original SOS program (Fig. 11.3). In this way, the details of the reformulation are abstracted from the user, who can work at the polynomial object level. The user interface of SOSTOOLS has been designed to be quite simple and easy to use, while maintaining a large degree of flexibility. The current version of SOSTOOLS uses either SeDuMi (Sturm, 1999) or SDPT3 (Toh et al., 1999), both of which are free MATLAB add-ons, as the SDP solver. The polynomial variables in the SOS programs can be defined in SOSTOOLS in two different ways: using the MATLAB Symbolic Math Toolbox or the custom-built polynomial toolbox. The former method provides the user the benefit of making use of all the features in the toolbox, which

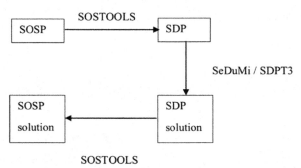

FIG. 11.3   Generic diagram depicting how SOS programs are solved using SOSTOOLS.

range from simple arithmetic operations to differentiation, integration and polynomial manipulation. Even though the integrated polynomial toolbox has only some of the functions of the symbolic toolbox, it allows users that do not have access to the symbolic toolbox to use SOSTOOLS. It also provides an alternative, sometimes faster SDP formulation path. In many cases the SDPs that we wish to solve have certain structural properties, such as sparsity, symmetry, etc. The formulation of the SDP should take them into account: this will not only reduce the computational burden of solving them as their size is many times reduced considerably, but it also removes numerical ill-conditioning.

When formulating the SDPs, provision has been taken for sparsity to be taken into account. The frequent use of certain SOS formulations, such as finding lower bounds on polynomial minima, and the search for Lyapunov functions for systems with polynomial vector fields are reflected in the introduction of customized functions in SOSTOOLS. A detailed description of how SOS-TOOLS works can be found in the SOSTOOLS user's guide (Prajna et al., 2002; Papachristodoulou and Prajna, 2002).

Lyapunov functions should be regarded as proofs guaranteeing the stability in chemical systems analysis. This aspect of Lyapunov methods is important, as analysis questions are addressed in a manner that no simulation procedure can. For a chemical system of the form $\dot{x} = f(x)$, assuming that function $f(x)$ is Lipschitz (which from a practical perspective implies it has a bounded first derivative), where $f(0) = 0$ (i.e., with the equilibrium assumed to be at the origin), it is stable if there exists a positive definite function $V(x)$ defined in some region of the state space containing the equilibrium point whose derivative $V(x) = \frac{\partial V}{\partial x} f(x)$ is negative semidefinite along the system trajectories. In the linear case, these conditions amount to finding, for a system $\dot{x} = Ax$ a positive definite matrix P such that $A^T P + PA$ is negative definite. Then the associated Lyapunov function is given by $V(x) = x^T Px$. Furthermore, for a multivariate polynomial $p(x)$, this is a sum of squares if there exist some polynomials $f_i(x)$, $i = 1, ..., M$ such that:

$$p(x) = \sum_{i=1}^{M} f_i^2(x) \tag{11.52}$$

This can be alternatively stated as follows: A polynomial $p(x)$ of degree $2d$ is an SOS if and only if there exists a positive semidefinite matrix Q and a vector of monomials $Z(x)$ containing monomials in x of degree $\leq d$ such that:

$$p(x) = Z(x)^T Q Z(x) \tag{11.53}$$

In general, the monomials in $Z(x)$ are not algebraically independent. Expanding $Z(x)^T Q Z(x)$ and equating the coefficients of the resulting monomials to the ones in $p(x)$, we obtain a set of affine relations in the elements of Q. Since $p(x)$ being SOS is equivalent to $Q \geq 0$, the problem of finding a Q, which proves that $p(x)$ is an SOS can be cast as a semidefinite program. Unfortunately, manually generated Lyapunov functions are inevitably limited to small-state dimensions and depend on the analytic skills of the researcher. In the case in which both the vector field f and the Lyapunov function candidate V are polynomial, the Lyapunov conditions are essentially polynomial nonnegativity conditions, which can be NP hard to test, and this has been one of the reasons for the lack of algorithmic constructions of Lyapunov functions. However, if we replace the nonnegativity conditions by the SOS conditions, then not only testing the Lyapunov function conditions—but also constructing the Lyapunov function—can be done efficiently using semidefinite programming (SDP). For example, as discussed in the example in Papachristodoulou and Prajna (2005), suppose that we want to know whether or not the quartic polynomial in two variables $p(x_1, x_2) = 2x_1^4 + 2x_1^3 x_2 - x_1^2 x_2^2 + 5x_2^4$ is a SOS, we can define $Z(x) = \left(x_1^2 \ x_2^2 \ x_1 x_2 \right)^T$ and consider the following quadratic form:

$$p(x_1, x_2) = Z(x)^T \underbrace{\begin{pmatrix} q_{11} & q_{12} & q_{13} \\ q_{12} & q_{22} & q_{23} \\ q_{13} & q_{23} & q_{33} \end{pmatrix}}_{Q} Z(x) \tag{11.54}$$

$$= q_{11} x_1^4 + q_{22} x_2^4 + (2q_{12} + q_{33}) x_1^2 x_2^2 + 2q_{13} x_1^3 x_2 + 2q_{23} x_1 x_2^3$$

from which one can obtain the following relations:

$q_{11} = 2$, $q_{22} = 5$, $q_{13} = 1$, $q_{23} = 0$, $2q_{12} + q_{33} = -1$. Now, decomposing $p(x)$ as an SOS amounts to searching for q_{12} and q_{33} satisfying the last equation, such that $Q \geq 0$. For $q_{12} = -3$ and $q_{33} = 5$, the matrix Q will be positive semidefinite and we have:

$$Q = L^T L, \quad L = \frac{1}{\sqrt{2}} \begin{pmatrix} 2 & -3 & 1 \\ 0 & 1 & 3 \end{pmatrix} \tag{11.55}$$

which leads to the following SOS decomposition:

$$p(x) = \frac{1}{2}\left(2x_1^2 - 3x_2^2 + x_1 x_2\right)^2 + \frac{1}{2}\left(x_2^2 + 3x_1 x_2\right)^2 \qquad (11.56)$$

Such Lyapunov functions can be found in a straightforward manner by using the *"findlyap"* command in SOSTOOLS.

Of further relevance to the design of controllers for chemical systems, the SOS approach also generalizes a well-known algorithmic tool in linear robust control theory—linear matrix inequalities (LMIs) (Boyd et al., 1994; Aylward et al., 2008) and enables the estimation of upper bounds of structured singular values (Dullerud and Paganini, 2000; Packard and Doyle, 1993). SOSTOOLS can thus be particularly useful in designs of bound-based extended Kalman filters (BEKF) (Hexner and Weiss, 2014). In the past, attempts at extended Kalman filtering for polynomial-based systems inevitably faced the closure problem (Jazwinski, 1970; Sorenson, 1974). The closure problem refers to the fact that to calculate the nth moment of a distribution, the value of the n + 1 and possibly higher-order moments are required. A popular method has been to assume that moments of higher order are related to lower-order moments as if the underlying probability density were Gaussian. For proper operation of any Kalman filter, it is essential that the calculated filter mean square error tracks the actual filter mean square error reasonably well. (As stated earlier, the reason for this is that the filter mean square error defines the filter gain.) Too small value of the calculated mean square error implies that the Kalman gain is too low and the observations are insufficiently weighted in updating the filter estimate. The formalization of this concept is called consistency (BarShalom et al., 2001). For a filter to be consistent, two conditions have to be fulfilled: (a) to have zero mean (unbiased) estimates and (b) to have covariance matrix as calculated by the filter. An important step in the development of such extended Kalman filters is the requirement for "tuning." This consists of adjusting (usually increasing) by trial and error the intensity of the process noise and possibly the observation noise so that the filter calculated mean square error is in some agreement with the actual mean square error. Using SOSTOOLS, it is possible to tune the filter for optimal performance according to these characteristics.

It is also worth noting that using Lyapunov-like arguments, other questions of relevance to the design of Kalman filters apart from stability can be answered algorithmically. For example, estimating L2 gains in nonlinear systems can be done by constructing appropriate storage functions (Willems, 1972a,b). Another application area is that of chemical reaction safety verification. In principle, safety verification or reachability analysis aims to show that starting at some initial conditions, a system cannot evolve to some unsafe region in the state space (Bemporad et al., 2000; Chutinan and Krogh, 2003; Papachristodoulou and Prajna, 2005).

Finally, other applications of the SOSTOOLS that can be explored in the context of chemometrics include the discrimination between rival biochemical network models (Mélykúti et al., 2010), the validation or invalidation of biological models (Anderson and Papachristodoulou, 2009), and as model decomposition and reduction tools for large-scale networks in systems biology (Anderson et al., 2011).

References

Abdeljawad, T., 2011. On Riemann and Caputo fractional differences. Comput. Math. Appl. 62, 1602–1611.

Águila, R.C., Carazo, A.H., Pérez, J.L., 2012. Extended and unscented filtering algorithms in nonlinear fractional order systems. Appl. Math. Sci. 6, 1471–1486.

Ahmad, H., Namerikawa, T., 2013. Extended Kalman filter-based mobile robot localization with intermittent measurements. Syst. Sci. Control Eng. 1, 113–126.

Ahn, C.K., 2012. Strictly passive FIR filtering for state-space models with external disturbance. Int. J. Electron. Commun. 66, 944–948.

Ahn, C.K., 2014. A new solution to the induced l∞ finite impulse response filtering problem based on two matrix inequalities. Int. J. Control. 87 (2), 404–409.

Ahn, C.K., Han, S., Kwon, W.H., 2006. H∞ FIR filters for linear continuous-time state-space systems. IEEE Signal Process. Lett. 13, 557–560.

Ahn, C.K., Shi, P., Basin, M.V., 2016. Deadbeat dissipative FIR filtering. IEEE Trans. Circuits Syst. I Regul. Pap. 63, 1210–1221.

Ait-El-Fquih, B., Desbouvries, F., 2006. Kalman filtering in triplet Markov chains. IEEE Trans. Signal Process. 54 (8), 2957–2963.

Anderson, J., Papachristodoulou, A., 2009. On validation and invalidation of biological models. BMC Bioinf. 10, 132.

Anderson, J., Yo-Cheng, C., Papachristodoulou, A., 2011. Model decomposition and reduction tools for large-scale networks in systems biology. Automatica 47, 1165–1174.

Applebaum, D., 2009. Lévy Processes and Stochastic Calculus, second ed. Cambridge University Press, Cambridge.

Ashayeri, L., Shafiee, M., Menhaj, M., 2013. Kalman filter for fractional order singular systems. J. Am. Sci. 9, 209–216.

Auger, F., Hilairet, M., Guerrero, J.M., Monmasson, E., 2013. Industrial applications of the Kalman filter: a review. IEEE Trans. Ind. Electron. 60, 5458–5471.

Aylward, E.M., Parriloa, P.A., Slotine, J.J.E., 2008. Stability and robustness analysis of nonlinear systems via contraction metrics and SOS programming. Automatica 44, 2163–2170.

Baccarelli, E., Cusani, R., 1996. Recursive Kalman-type optimal estimation and detection of hidden Markov chains. Signal Process. 51, 55–64.

BarShalom, Y., Li, X.R., Kirubajan, T., 2001. Estimation With Applications to Tracking and Navigation: Theory Algorithm and Software. John Wiley and Sons.

Basseville, M., Benveniste, A., Chou, K.C., Golden, S.A., Nikoukhah, R., Willsky, A.S., 1992. Modeling and estimation of multiresolution stochastic processes. IEEE Trans. Inf. Theory 38, 766–784.

Bastin, G., Dochain, D., 1990. On-Line Estimation and Adaptive Control of Bioreactors. Elsevier, Amsterdam.

Becis-Aubry, Y., Boutayeb, M., Darouach, M., 2008. State estimation in the presence of bounded disturbances. Automatica 44, 1867–1873.

Bemporad, A., Torrisi, F.D., Morari, M., 2000. Optimization-based verification and stability characterization of piecewise affine and hybrid systems. In: Hybrid Systems: Computation and Control, LNCS 1790. Springer-Verlag, pp. 45–58.

Betancur, M.J., Moreno, J.A., Moreno-Andrade, I., Buitrón, G., 2006. Practical optimal control of fed-batch bioreactors for the waste water treatment. Int. J. Robust Nonlinear Control 3, 173–190.

Bettayeb, M., Mansouri, R., 2014. IMC-PID-fractional-order-filter controllers design for integer order systems. ISA Trans. 53, 1620–1628.

Bogaerts, P., 1999. A hybrid asymptotic Kalman observer for bioprocesses. Bioprocess Eng. 20, 101–113.

Boizot, N., Busvelle, E., Gautier, J.P., 2010. An adaptive high gain observer for nonlinear systems. Automatica 46, 1483–1488.

Bonnabel, S.S., Martin, P., Salaün, E., 2009. Invariant extended Kalman filter: theory and application to a velocity-aided attitude estimation problem. In: Proceedings of the 48th IEEE Conference on Decision and Control. pp. 1297–1304.

Bourmaud, G., Mégret, R., Giremus, A., Berthoumieu, Y., 2016. From intrinsic optimization to iterated extended Kalman filtering on Lie groups. J. Math. Imaging Vision 55, 284–303.

Boyd, S., El Ghaoui, L., Feron, E., Balakrishnan, V., 1994. Linear Matrix Inequalities in System and Control Theory. Society for Industrial and Applied Mathematics (SIAM).

Brown, R.G., Patrick, Y., Hwang, C., 1997. Introduction to Random Signals and Applied Kalman Filtering: With Matlab Exercises and Solutions, third ed. John Wiley & Sons, New York.

Cao, J., Cao, B., 2006. Design of fractional order controllers based on particle swarm optimization. Int. J. Control. Autom. Syst. 4, 775–781.

Caponetto, R., Dongola, G., 2013. A numerical approach for computing stability region of FO-PID controller. J. Franklin Inst. 350, 871–889.

Carli, R., Chiuso, A., Schenato, L., Zampieri, S., 2008. Distributed Kalman filtering based on consensus strategies. IEEE J. Sel. Areas Commun. 26, 622–633.

Carmi, A., Gurfil, P., Kanevsky, D., 2010. Methods for sparse signal recovery using Kalman filtering with embedded pseudo-measurement norms and quasi-norms. IEEE Trans. Signal Process. 58, 2405–2409.

Chutinan, A., Krogh, B.H., 2003. Computational techniques for hybrid system verification. IEEE Trans. Autom. Control 48, 64–75.

Chyi-Tsong, C., Shih-Tien, P., 2005. Design of a sliding mode control for systems for chemical process. J. Process Control 15, 515–530.

Cox, H., 1964. On the estimation of state variables and parameters for noisy dynamic systems. IEEE Trans. Autom. Control 9, 5–12.

Cruz-Victoria, J., Martinez-Guerra, R., Rincón-Pasaye, J., 2008. On nonlinear systems diagnosis using differential and algebraic methods. J. Franklin Inst. 345, 102–118.

Dadras, S., Dadras, S., Malek, H., Chen, Y.Q., 2017. A note on the Lyapunov stability of fractional-order nonlinear systems. In: Proceedings of the ASME 2017 International Design Engineering Technical Conferences & Computers and Information in Engineering Conference. Paper No. IDETC2017-68270.

Daum, F., 2005. Nonlinear filters: beyond the Kalman filter. IEEE Aerosp. Electron. Syst. Mag. 20, 5769.

Diop, S., 2001. The algebraic theory of nonlinear observability revisited. In: Proceedings of the 40th IEEE Conference on Decision and Control, Florida, USA. pp. 2550–2555.

Diop, S., Filess, M., 1991. Nonlinear observability identifiability and persistent trajectories. In: Proceedings of the 30th IEEE Conference on Decision and Control, Brighton, England. pp. 714–719.

Dullerud, G.E., Paganini, F., 2000. A Course in Robust Control Theory: A Convex Approach. Springer-Verlag, New York.

Efe, M.O., 2011. Fractional order systems industrial automation—a survey. IEEE Trans. Ind. Inf. 7, 582–591.

El-Sayed, A.M.A., 1996. Fractional-order diffusion-wave equation. Int. J. Theor. Phys. 35, 311–322.

Evensen, G., 1994. Sequential data assimilation with nonlinear quasi-geostrophic model using Monte Carlo methods to forecast error statistics. J. Geophys. Res. 99, 143–162.

Fliess, M., 1990. Generalized controller canonical forms for linear and nonlinear systems. IEEE Trans. Autom. Control 35, 994–1008.

Fliess, M., Lévine, J., Martín, P., Rouchón, P., 1995. Flatness and defect of nonlinear systems: introductory theory and examples. Int. J. Control. 61, 1327–1361.

Gao, Z., 2018. Fractional-order Kalman filters for continuous-time fractional-order systems involving colored process and measurement noises. J. Franklin Inst. 355, 922–948.

Gauthier, J.P., Hammouri, H., Othman, S., 1992a. Observability for any u(t) of a class of bilinear systems. IEEE Trans. Autom. Control 37, 875–880.

Gauthier, J.P., Hammouri, H., Othman, S., 1992b. A simple observer for nonlinear systems—applications to bioreactors. IEEE Trans. Autom. Control 37, 875–880.

Gibbs, B., 2011. Advanced Kalman Filtering, Least-Squares and Modeling. Wiley, New York.

Gouze, J.L., Rapaport, A., Hadj-Sadok, M.Z., 2000. Interval observers for uncertain biological systems. Ecol. Model. 133, 45–56.

Guan, X., Peng, H., Li, L., Wang, Y., 2001. Parameter identification and control of Lorenz chaotic system. Acta Phys. Sin. 50, 26–29.

Hamamci, S.E., 2007. An algorithm for stabilization of fractional-order time delay systems using fractional-order PID controllers. IEEE Trans. Autom. Control 52, 1964–1969.

Han, S.H., Kwon, W.H., Kim, P.S., 2002. Quasi-deadbeat minimax filters for deterministic state-space models. IEEE Trans. Autom. Control 47, 1904–1908.

Hanlon, P.D., Maybeck, P.S., 2000. Characterization of Kalman filter residuals in the presence of mismodeling. IEEE Trans. Aerosp. Electron. Syst. 36, 114–131.

Hexner, G., Weiss, H., 2014. An extended Kalman filter with a computed mean square error bound. In: 53rd IEEE Conference on Decision and Control December 15–17, 2014, Los Angeles, CA, USA.

Hilfer, R., 2000. Applications of Fractional Calculus in Physics. World Scientific, River Edge, New Jersey.

Hu, J., Wang, Z., Shen, B., Gao, H., 2013. Quantised recursive filtering for a class of nonlinear systems with multiplicative noises and missing measurements. Int. J. Control. 86, 650–663.

Isidori, A., 1999. Nonlinear Control Systems. Springer, New York.

Jazwinski, A.H., 1968. Limited memory optimal filtering. IEEE Trans. Autom. Control 13, 558–563.

Jazwinski, A.H., 1970. Stochastic Processes and Filtering Theory. Academic Press, New York.

Julier, S.J., Uhlmann, J.K., 1997. A new extension of the Kalman filter to nonlinear systems. Proc. SPIE 3068, 182–193.

Kalman, R.E., 1960. A new approach to linear filtering and prediction problems. J. Fluids Eng. 82, 35–45.

Keller, H., 1987. Non-linear observer design by transformation into a generalized observer canonical form. Int. J. Control. 46, 1915–1930.

Khalil, H., 2002. Nonlinear Systems, third ed. Prentice Hall, New Jersey.

Kim, P.S., 2010. An alternative FIR filter for state estimation in discrete-time systems. Digital Signal Process. 20, 935–943.

Kim, P.S., Lee, M.E., 2007. A new FIR filter for state estimation and its applications. J. Comput. Sci. Technol. 22, 779–784.

Kucera, V., 1991. Analysis and Design of Discrete Linear Control Systems. Prentice Hall, New York.

Kwon, W.H., Han, S., 2005. Receding Horizon Control: Model Predictive Control for State Models. Springer-Verlag, London.

Kwon, O.K., Kwon, W.H., Lee, K.S., 1989. FIR filters and recursive forms for discrete-time state-space models. Automatica 25, 715–728.

Kwon, W.H., Kim, P.S., Han, S.H., 2002. A receding horizon unbiased FIR filter for discrete-time state space models. Automatica 38, 545–551.

Lange, A.A., 2001. Simultaneous statistical calibration of the GPS signal delay measurements with related meteorological data. Phys. Chem. Earth Solid Earth Geod. 26, 471–473.

Li, Y., Chen, Y.Q., Ahn, H.S., 2011. Fractional-order iterative learning control for fractional-order linear systems. Asian J. Control 13, 54–63.

Liang, J., Wang, Z., Liu, Y., 2014. State estimation for two-dimensional complex networks with randomly occurring nonlinearities and randomly varying sensor delays. Int. J. Robust Nonlinear Control 24, 18–38.

Liao, Z., Peng, C., Li, W., Wang, Y., 2011. Robust stability analysis for a class of fractional order systems with uncertain parameters. J. Franklin Inst. 348, 1101–1113.

Luenberger, D.G., 1964. Observing the state of a linear system. IEEE Trans. Mil. Electron. 8, 74–80.

Luenberger, D., 1971. An introduction to observers. IEEE Trans. Autom. Control 16, 592–602.

Machado, J.A.T., Galhano, A.M.S.F., Trujillo, J.J., 2014. On development of fractional calculus during the last fifty years. Scientometrics 98, 577–582.

Maronna, R.A., Yohai, V.J., 2006. Robust Statistics Theory and Methods. Wiley, New York.

Martin, R.D., Masreliez, C.J., 1975. Robust estimation via stochastic approximation. IEEE Trans. Inf. Theory IT-21, 263–271.

Martínez-Guerra, R., Diop, S., 2004. Diagnosis of nonlinear systems: an algebraic and differential approach. IEE Proc. Control Theory Appl. 151, 130–135.

Masreliez, C.J., 1975. Approximate non-Gaussian filtering with linear state and observation relations. IEEE Trans. Autom. Control AC-20, 107–110.

Mélykúti, B., August, E., Papachristodoulou, A., El-Samad, H., 2010. Discriminating between rival biochemical network models: three approaches to optimal experiment design. BMC Syst. Biol. 4, 38.

Meta-Machuca, J.L., Martínez-Guerra, R., Aguilar-López, R., Gray, P., Scott, S.K., 1990. Chemical Oscillations and Instabilities. Clarendon Press, Oxford.

Monje, C., Chen, Y., Vinagre, B., Xue, D., et al., 2010. Fractional-order systems and controls: fundamentals and applications. In: Series on Advances in Industrial Control, Springer, London, UK.

Nahi, N.E., 1969. Optimal recursive estimation with uncertain observation. IEEE Trans. Inf. Theory IT-15, 457–462.

Najar, S., Abdelkrim, M.N., Abdelhamid, M., et al., 2009. Discrete fractional Kalman filter. IFAC Proc. 42, 520–525.

Odlham, K.B., Spaniar, J., 1974. The Fractional Calculus: Theory and Applications of Differentiation and Integration to Arbitrary Order. Academic Press, New York.

Packard, A., Doyle, J.C., 1993. The complex structured singular value. Automatica 29 (1), 71–109.

Papachristodoulou, A., Prajna, S., 2002. On the construction of Lyapunov functions using the sum of squares decomposition. In: Proceedings of the 41st IEEE Conference on Decision and Control Las Vega, Nevada, USA. pp. 3482–3487, December.

Papachristodoulou, A., Prajna, S., 2005. A tutorial on sum of squares techniques for systems analysis. In: American Control Conference June 8–10, 2005, Portland, OR, USA, pp. 2686–2700.

Papachristodoulou, A., Anderson, J., Valmorbida, G., Prajna, S., Seiler, P., Parrilo, P., 2013. SOSTOOLS Version 3.00 Sum of Squares Optimization Toolbox for MATLAB. arXiv:1310.4716 [math.OC].

Parrilo, P.A., 2000. Structured Semidefinite Programs and Semialgebraic Geometry Methods in Robustness and Optimization. PhD thesis, California Institute of Technology, Pasadena, CA.

Parrilo, P.A., Lall, S., 2003. Semidefinite programming relaxations and algebraic optimization in control. Eur. J. Control. 9, 307–321.

Parrilo, P.A., Sturmfels, B., 2001. Minimizing polynomial functions. In: Workshop on Algorithmic and Quantitative Aspects of Real Algebraic Geometry in Mathematics and Computer Science.

Podlubny, I., 1999. Fractional Differential Equations. Academic Press, New York.

Pourdehi, S., Azami, A., Shabaninia, F., 2015. Fuzzy Kalman-type filter for interval fractional-order systems with finite-step auto-correlated process noises. Neurocomputing 159, 44–49.

Prajna, S., Papachristodoulou, A., Parrilo, P.A., 2002. SOSTOOLS—Sum of Squares Optimization Toolbox, User's Guide. Available at http://www.cds.caltech.edu/sostools.

Rakkiyappan, R., Zhu, Q., Chandrasekar, A., 2014. Stability of stochastic neural networks of neutral type with Markovian jumping parameters: a delay-fractioning approach. J. Franklin Inst. 351, 1553–1570.

Rawicz, P.L., Kalata, P.R., Murphy, K.M., Chmielewski, T.A., 2003. Explicit formulas for two state Kalman, H2 and H∞ target tracking filters. IEEE Trans. Aerosp. Electron. Syst. 39, 53–69.

Rong, X., Bar-Shalom, Y., 1994. A recursive multiple model approach to noise identification. IEEE Trans. Aerosp. Electron. Syst. 30, 671–684.

Sadeghian, H., Salarieh, H., 2013. On the general Kalman filter for discrete time stochastic fractional systems. Mechatronics 23, 764–771.

Safarinejadian, B., Kianpour, N., Asad, M., 2018. State estimation in fractional-order systems with coloured measurement noise. Trans. Inst. Meas. Control. 40, 1819–1835. https://doi.org/10.1177/0142331217691219.

Sarkka, S., Nummenmaa, A., 2009. Recursive noise adaptive Kalman filtering by variational Bayesian approximation. IEEE Trans. Autom. Control 54, 596–600.

Shen, B., Wang, Z., Shu, H., 2008. On nonlinear H1 filtering for discrete-time stochastic systems with missing measurements. IEEE Trans. Autom. Control 53, 2170–2180.

Shmaliy, Y.S., 2009. GPS-Based Optimal FIR Filtering of Clock Models. Nova Science, New York.

Shmaliy, Y.S., 2011. An iterative Kalman-like algorithm ignoring noise and initial conditions. IEEE Trans. Signal Process. 59, 2465–2473.

Sierociuk, D., 2013. Fractional Kalman filter algorithms for correlated system and measurement noises. Control. Cybern. 42 (2), 471–490.

Sierociuk, D., Dzieliński, A., 2006. Fractional Kalman filter algorithm for the states, parameters and order of fractional system estimation. Int. J. Appl. Math. Comput. Sci. 16, 129–140.

Sierociuk, D., Ziubinski, P., 2014. Fractional order estimation schemes for fractional and integer order systems with constant and variable fractional order colored noise. Circ. Syst. Signal Process. 33, 3861–3882.

Sierociuk, D., Tejado, I., Vinagre, B.M., 2011. Improved fractional Kalman filter and its application to estimation over lossy networks. Signal Process. 91, 542–552.

Sierociuk, D., Macias, M., Malesza, W., et al., 2016. Dual estimation of fractional variable order based on the unscented fractional order Kalman filter for direct and networked measurements. Circ. Syst. Signal Process. 35, 2055–2082.

Simon, D., 2006. Optimal State Estimation: Kalman, H∞, and Nonlinear Approaches. Wiley, Hoboken, NJ.

Simon, D., 2010. Kalman filtering with state constraints: a survey of linear and nonlinear algorithms. IET Control Theory Appl. 4, 1303–1318.

Simon, D., Shmaliy, Y.S., 2013. Unified forms for Kalman and finite impulse response filtering and smoothing. Automatica 49, 1892–1899.

Sorenson, H.W., 1974. On the development of practical nonlinear filters. Inform. Sci. 7, 253–270.

Sorenson, S.W., Sacks, J.E., 1971. Recursive fading memory filters. Inform. Sci. 3, 101–119.

Sturm, J.F., 1999. Using SeDuMi 1.02, a Matlab toolbox for optimization over symmetric cones. Optim. Methods Softw. 11–12, 625–653. Available at http://fewcal.kub.nl/sturm/software/sedumi.html.

Sun, X., Duan, J., Li, X., Wang, X., 2013a. State estimation under non-Gaussian Levy noise: a modified Kalman filtering method. In: Proceedings of the Banach Center.arXiv:1303.2395.

Sun, X., Yan, G., Zhang, B., 2013b. Kalman smoother and weighted fusion algorithm for fractional systems. In: International Conference on Measurement, Information and Control. vol. 1. pp. 151–158, August.

Toh, K.C., Tütüncü, R.H., Todd, M.J., 1999. SDPT3—a Matlab software package for semi-definite quadratic-linear programming. Optim. Methods Softw. 11, 545–581. Available at http://www.math.nus.edu.sg/mattohkc/sdpt3.html.

Torabi, H., Pariz, N., Karimpour, A., 2016. Kalman filters for fractional discrete-time stochastic systems along with time-delay in the observation signal. Eur. Phys. J. Spec. Top. 225, 107–118.

Vandenberghe, L., Boyd, S., 1996. Semidefinite programming. SIAM Rev. 38 (1), 49–95.

Vaswani, N., 2008. Kalman filtered compressed sensing. In: Proceedings of the 15th IEEE International Conference on Image Processing. pp. 893–896.

Vinagre, B.M., Chen, Y.Q., Petras, I., 2003. Two direct Tustin discretization methods for fractional-order differentiator/integrator. J. Franklin Inst. 340 (5), 349–362.

Wang, Z., Huang, X., Zhou, J., 2013. A numerical method for delayed fractional-order differential equations: based on G-L definition. Appl. Math. Inf. Sci. 7, 525–529.

Wang, H., Yu, Y., Wen, G., 2014. Stability analysis of fractional-order Hopfield neural networks with time delays. Neural Netw. 55, 98–109.

Wei, G., Wang, Z., Shu, H., 2009. Robust filtering with stochastic nonlinearities and multiple missing measurements. Automatica 45, 836–841.

Willems, J.C., 1972a. Dissipative dynamical systems. I. General theory. Arch. Ration. Mech. Anal. 45, 321–351.

Willems, J.C., 1972b. Dissipative dynamical systems. II. Linear systems with quadratic supply rates. Arch. Ration. Mech. Anal. 45, 352–393.

Wu, X., Sun, Y., Lu, Z., Wei, Z., Ni, M., Yu, W., 2015. Modified Kalman filter algorithm for fractional system under Lévy noises. J. Franklin Inst. 352, 1963–1978.

Yang, Z., Cao, J., 2013. Initial value problems for arbitrary order fractional differential equations with delay. Commun. Nonlinear Sci. Numer. Simul. 18, 2993–3005.

Yin, C., Chen, Y.Q., Zhong, S.M., 2014. Fractional-order sliding mode based extremum seeking control of a class of nonlinear systems. Automatica 50, 3173–3181.

Yin, C., Cheng, Y., Chen, Y.Q., et al., 2015. Adaptive fractional-order switching-type control method design for 3D fractional-order nonlinear systems. Nonlinear Dyn. 82, 39–52.

Yin, C., Huang, X., Chen, Y.Q., et al., 2017. Fractional-order exponential switching technique to enhance sliding mode control. Appl. Math. Model. 44, 705–726.

Zhang, L., et al., 2014. H1 filtering for a class of discrete-time switched fuzzy systems. Nonlinear Anal. Hybrid Syst. 14, 74–85.

Further reading

Azami, A., Naghavi, S.V., Tehrani, R.D., et al., 2017. State estimation strategy for fractional order systems with noises and multiple time delayed measurements. IET Sci. Meas. Technol. 11, 9–17.

Caballero-Aguila, R., Hermoso-Carazo, A., Linares-Perez, J., 2012. Extended and unscented filtering algorithms in nonlinear fractional order systems with uncertain observations. Appl. Math. Sci. 6, 1471–1486.

Sun, X.W.Y., Lu, Z., 2015. A modified Kalman filter algorithm for fractional system under Lévy noises. J. Franklin Inst. 352, 1963–1978.

Chapter 12

Recent advances in chemometrics

Chapter outline

12.1 Recent advances in set membership approaches to state estimation and mixture analysis techniques

12.1.1 Traditional chemometric approaches to mixture analysis

As discussed in the previous chapters, multivariate techniques are often adopted in chemometrics to perform mixture analysis. It is possible to perform multivariate curve resolution using several techniques. The most straightforward way is

State Estimation in Chemometrics. https://doi.org/10.1016/B978-0-08-102603-8.00012-2
213

through PCA with rotation. An alternative is to add constraints using the geometrical properties of the multivariate space (Borgen and Kowalski, 1985). In addition, one can try to find pure component regions using interactive self-modeling mixture analysis as discussed in Windig and Guilment (1991), or adopt an orthogonal projection approach (Cuesta-Sanchez et al., 1996; Vandeginste and Derks, 1985). Further to the well-established least squares techniques that have been discussed in many textbooks in chemometrics (Martens and Næcs, 1989; Adams, 1995; Beebe et al., 1998; Kemsley, 1998; Martens and Martens, 2000; Brereton, 2003), constrained least square algorithms may be used, e.g., Karjalainen and Karjalainen (1996). These constrained least squares algorithms often are implemented using linear inequality constraints. They have been discussed in the context of high-performance liquid chromatography (HPLC) (Vandeginste and Derks, 1985), spectroscopic-chromatographic data (Gemperline, 1986), or used in multivariate curve resolution for liquid chromatography using diode array detection (Tauler and Barcelo, 1993).

In addition, local rank-based techniques may be adopted (Maeder, 1987; Liang et al., 1993) as a means to perform evolving factor analysis for the resolution of overlapping chromatographic peaks. Finally, there are rank annihilation techniques for factor analysis as discussed by Ho et al. (1978), and Sanchez and Kowalski (1986).

For multivariate calibration, there are ordinary least squares (OLS)/multiple linear regression (MLR) techniques (Draper and Smith, 1981), Ridge regression (RR) techniques (Hoerl and Kennard, 1970), or one can use principal component regression (PCR) (Massy, 1965), latent root regression (LRR) (Webster et al., 1974), partial least squares regression (PLS) (Helland, 1988; Hoskuldsson, 1988), sliced inverse regression (SIR) (Li, 1991), continuum regression (CR) (Stone and Brooks, 1990), locally weighted regression (LWR) (Næcs and Isaksson, 1989), and principal covariates regression (PCovR) approaches (de Jong and Kiers, 1992). When the relations are nonlinear, alternative algorithms may be used (Geladi and Dábakk, 1995; Geladi et al., 1985; Barnes et al., 1989; Geladi, 2001; Svensson et al., 2002).

12.1.2 State space approaches to mixture analysis

Most of the aforementioned algorithms have been applied in a postprocessing stage after data acquisition. The current challenge in mixture analysis for process control lies in estimating time-evolving states while incorporating fixed model parameters in a state-space model context. Chen and Liu (2000) place this estimation task in a Mixture Kalman Filters context. This is particularly important when more complex processes are observed, e.g., in synthetic bioengineering (Tan et al., 2007), in fluorescent microscopy (Wang et al., 2010). They are also particularly relevant to gene regulation studies (Rosenfeld et al., 2005; Wilkinson, 2006) and in Bayesian inference of stochastic kinetic models (Golightly and Wilkinson, 2005) as high-resolution time series observations are becoming experimentally possible.

As explained in the article by Niemi and West (2010), given a specified model and a series of observations over a fixed time interval, we are interested in summaries of the posterior distribution for the full series of corresponding state vectors as well as fixed model parameters; this is the batch analysis. Application of sequential filtering and retrospective smoothing using Monte Carlo is used for the nonlinear forward filtering process. A backward sampling (FFBS) approach can be used.

Several Markov chain Monte Carlo (MCMC) approaches have been suggested in the literature to develop nonlinear non-Gaussian state-space models; these include the works of Carlin et al. (1992), and Geweke and Tanizaki (1999). In addition, Stroud et al. (2003) suggested sampling latent states in blocks from an auxiliary mixture model. The main difference to the method discussed by Niemi and West (2010) is that their MCMC scheme does not condition on mixture component indicators.

Another relevant work on dynamic generalized linear models to that of Niemi and West (2010) is that of West and Harrison (1997). They developed a global Metropolis-Hastings analysis where the proposal distribution for state vectors is generated from analytic approximations to filtering and smoothing distributions that are known to be accurate. The reader is referred to the article by Niemi and West (2010) and the web page http://ftp.stat.duke.edu/WorkingPapers/08-21.html, which provides freely available MATLAB code that implements the MCMC algorithm discussed to perform adaptive mixture modeling of nonlinear state space models.

12.1.3 Set-membership approaches for state estimation

For state estimation, it is widely accepted that it is more realistic to assume that the perturbations and measurement noises are unknown but bounded, so the task becomes finding a methodology for guaranteed state estimation as firstly introduced by Schweppe (1968) and Wistenhausen (1968). Set-membership approaches have been developed by Bertsekas and Rhodes (1971), as well as by Fogel and Huang (1982). In these approaches, the idea is to compute a compact set that guarantees to contain the set of states that are consistent with the model of the system, the realization of the measurement and the bounded perturbations and measurement noises. To implement set-membership estimation techniques, several sets may be used. Kurzhanski and Vályi (1996), Durieu et al. (2001), Polyak et al. (2004), Daryin et al. (2006), Daryin and Kurzhanski (2012), as well as Chernousko (1994) discussed the use of ellipsoids. Other approaches such as the one by Walter and Piet-Lahanier (1989) included the use of polytopes, whereas Chisci et al. (1996) discussed the use of parallelotopes. Furthermore, Puig et al. (2001), Combastel (2003), Alamo et al. (2005), and Le et al. (2013) discussed the use of zonotopes. Ellipsoids are more widely used, however, due to the simplicity of their formulation and resulting estimation stability properties (Hu and Lin, 2003). As discussed in Durieu et al. (2001), to minimize the size of the estimation ellipsoidal set, two methods are mainly considered. Firstly, the determinant-based criterion is minimized, which is equivalent to minimize the volume of the ellipsoidal set. Secondly, the minimization of the trace criterion, which is equivalent to minimize the sum of squares of the half length of the axes of the ellipsoid, is considered. These two methods offer low complexity, but with a loss of accuracy compared to the polytopic estimation. A recent advance in this respect is the work by Chabane et al. (2014), which proposes a new ellipsoid-based guaranteed state estimation approach where state estimation is computed by casting it as an offline Linear Matrix Inequality optimization problem, which aims to minimize the radius of the ellipsoidal estimation set.

12.2 Recent advances in wavelet methods in chemometrics

12.2.1 Chemometrics in the context of parsimonious representation of signals

Linear transforms are useful both for noise extraction and for representing the information in the data using fewer coefficients in a parsimonious manner. Noise extraction can be performed by assuming that the system is detector noise limited rather than source noise limited (e.g., sufficient stability in amplitude and phase for a laser source used for the measurements, etc.). In this case, the noise spreads equally among all transform coefficients (white Gaussian), while useful information will generally be concentrated in fewer coefficients. Examples of alternative linear transformations to the PCA that was discussed earlier in the book are the Fourier transform (FT), the windowed FT, and wavelet transforms (WT). These have been used extensively within the context of chemometrics for the processing of spectroscopic data. The main aim of PCA is maximum information compression. A disadvantage of this transform is that each PC is related to the whole spectrum, and no a priori assumption on a particular model is associated with the dataset structure. Thus, "problematic" regions in the spectrum cannot be excluded after the computation of the PCs. Furthermore, since the analysis functions are obtained from the statistics of the data, many calibration samples may be required to obtain reliable PCs. In contrast, in the case of Fourier and wavelet transforms, the analysis functions are fixed. And in the case of Fourier transforms, spatial information is lost. As a result, problematic regions in the spectrum cannot be excluded after the transform. Compacting a waveform using Fourier coefficients is not as efficient when samples have complex spectra and filtering too many Fourier coefficients can lead to signal distortion. These issues are partly addressed with the adoption of windowed Fourier transforms, and eliminated with the use of wavelet transforms. Wavelet transforms are particularly useful in this respect, as spatial information is kept. Furthermore, the width of the analysis window is automatically varied. For best performance, adaptive wavelets should be considered, but the optimization methodology can have significant impact on classifier performance if the transform is used as part of a preprocessing step for denoising and feature extraction. A comprehensive review of linear transforms within a chemometrics context that provides a comparison of all these transforms and discusses filtering and regression can be found in the work by Hadjiloucas et al. (2002).

The benefits of wavelet transform in a chemometrics context can be observed by looking at single and coaveraged interferograms obtained using a continuous-wave Fourier transform spectrometer operating at the THz part of the spectrum in Fig. 12.1. As it can be observed, individual interferograms

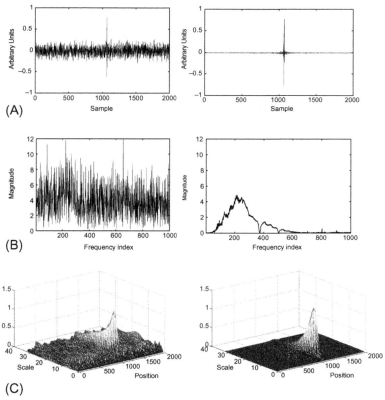

FIG. 12.1   (A) Time-domain, (B) Fourier transformed and (C) wavelet transformed individual (noisy) and coaveraged interferograms from a continuous-wave rapid scan Fourier transform THz spectrometer. *(Results reproduced from Hadjiloucas, S., Galvão, R.K.H., Bowen J.W., 2002. Analysis of spectroscopic measurements of leaf water content at THz frequencies using linear transforms, J. Opt. Soc. Am. A 19, 2495–2509.)*

can be too noisy and their corresponding Fourier transform spectra can look even more noisy, yet a wavelet transform of the dataset can still provide a useful signal because the energy in the noise is spread across an entire plane in the case of the wavelet transform as opposed to a single dimension in the case of the Fourier transform. Furthermore, the coaveraging of interferograms improves the signal-to-noise ratio by a factor equal to the square root of the number of coaveraged interferograms.

Clearly, from Fig. 12.1 it can be observed that wavelet transformation of a dataset can be particularly useful and can enable feature extraction even under conditions of low signal-to-noise ratio. This is particularly important in a chemometrics context when measurements are noisy or need to be performed within short time scales. Wavelet transformation, with filtering of the wavelet coefficients and reconstruction of the original waveform, is shown qualitatively in Fig. 12.2.

Filtering may be performed using a filter bank formulation of the fast wavelet transform (Strang and Nguyen, 1998). Each time domain signature is represented by a data vector of length J, and a change in variables is performed in transform space $T: \mathfrak{R}^J \rightarrow \mathfrak{R}^J$, $T(\mathbf{x}) = \mathbf{t}$, with the elements of $\mathbf{t}$ given by.

$$t_k = \sum_{n=0}^{J-1} x_n v_k(n), \quad k=0,1,\ldots,J-1, \tag{12.1}$$

where $v_k(n) \in \mathfrak{R}$ is a transform weight. It proves convenient to write the transform in matrix form as:

$$\mathbf{t}_{1\times J} = \mathbf{x}_{1\times J} \mathbf{V}_{J\times J}, \tag{12.2}$$

where $\mathbf{x}$ is the row vector of original variables, $\mathbf{t}$ is the row vector of new (transformed) variables, and $\mathbf{V}$ is the matrix of weights. Choosing $\mathbf{V}$ to be unitary (that is, $\mathbf{V}^T\mathbf{V} = \mathbf{I}$), the transform is said to be orthogonal and it therefore consists of a

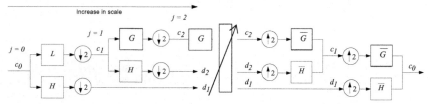

FIG. 12.2 Wavelet decomposition, filtering in the wavelet domain and wavelet reconstruction with the inverse wavelet transform. Two-channel filter bank. Blocks H and G represent a lowpass and a highpass filter, respectively, and $\downarrow 2$ denotes the operation of dyadic downsampling and the overbar is used to denote filtered response from each filter bank. The decomposition can be carried out in more resolution levels by successively splitting the lowpass channel. The larger arrow denotes filtering across all bands.

simple rotation in the coordinate axes (with the new axes directions determined by the columns of $\mathbf{V}$).

As discussed in the wavelet literature (Daubechies, 1992; Vaidyanathan, 1993; Sherlock and Monro, 1998), the discrete wavelet transform can be calculated in a fast manner by using a filter bank structure of the form depicted in Fig. 12.2. In this filter bank, the low-pass filtering result undergoes successive filtering iterations with the number of iterations N ($N < J$) chosen by the analyst. The final result of the decomposition of data vector $\mathbf{x}$ is a vector resulting from the concatenation of row vectors $\mathbf{c}(N)$ and $\mathbf{d}(N)$ (termed approximation and detail coefficients at the Nth scale level, respectively) in the following manner:

$$\mathbf{t} = [\mathbf{c}(N) \mid \mathbf{d}(N) \mid \mathbf{d}(N-1) \mid \ldots \mid \mathbf{d}(1)] \tag{12.3}$$

with coefficients in larger scales (e.g., $\mathbf{d}(1),\mathbf{d}(2),\mathbf{d}(3)$, ...) associated to broad features in the data vector, and coefficients in smaller scales (e.g., $\mathbf{d}(N)$, $\mathbf{d}(N\text{-}1),\mathbf{d}(N\text{-}2)\ldots$) associated to narrower features such as sharp peaks.

Since employing the same filters in all levels of the bank is not a strict requirement, a more general filtering and down-sampling procedure for a data vector of J points $\mathbf{x} = [x_0\ x_1\ \ldots\ x_J]$ with $2\,N$ levels of low-pass and $2\,N$ levels of high-pass filtering sequences $\{H_0, H_1, H_2, \ldots, H_{2N\text{-}1}\}$ and $\{G_0, G_1, G_2, \ldots, G_{2N\text{-}1}\}$ may be used, with impulse responses $\{h_0, h_1, \ldots, h_{2N\text{-}1}\}$ and $\{g_0, g_1, \ldots, g_{2N\text{-}1}\}$, respectively. After the data vector is periodized to minimize border effects, the convolution consists of flipping the filtering sequence and moving it alongside the data vector as shown in Table 12.1. For generating the low-pass filter coefficients, for each position of the filtering sequence with respect to the data vector, the scalar product of the two is calculated (with missing points in the filtering sequence replaced with zeros). Table 12.1 depicts this process for the low-pass filter coefficients, e.g., $c_1' = x_1h_3 + x_2h_2 + x_3h_1 + x_4h_0$. Dyadic downsampling is then performed to c_{2i}' to generate coefficients c_i. A similar table can be constructed

TABLE 12.1 Convolution and downsampling procedure for $2\,N$ level low-pass filtering.

x_0	x_1	x_2	x_3	...	x_J	x_0	x_1		Before	After
h_{2N-1}	...	h_1	h_0	...					c_0'	
	h_{2N-1}	...	h_1	h_0	...				c_1'	c_0
		h_{2N-1}	...	h_1	h_0		...		c_2'	
			h_{2N-1}	...	h_1	h_0	...		c_3'	c_1
				⋮	⋮	⋮	⋮	...	⋮	
					h_{2N-1}	...	h_1		c_{2N-1}'	c_{N-1}

for the calculation of high-pass coefficients with the downsampling operator applied to d_{2i}' to generate coefficients d_i.

In the general case, for a data vector $\mathbf{x}$ of length J and filtering sequences of length $2N$ each, the transformation matrix resulting from the concatenation of both high $\mathbf{G}$ and low $\mathbf{H}$ frequency responses becomes $\mathbf{V} = [\mathbf{H} \mid \mathbf{G}]$, or.

$$
\mathbf{V} =
\begin{bmatrix}
0 & 0 & \cdots & h_{2N-4} & h_{2N-2} & 0 & 0 & \cdots & g_{1N-4} & g_{1N-2} \\
h_{2N-1} & 0 & \cdots & h_{2N-5} & h_{2N-3} & g_{1N-1} & 0 & \cdots & g_{1N-5} & g_{1N-3} \\
h_{2N-2} & 0 & \cdots & h_{2N-6} & h_{2N-4} & g_{1N-2} & 0 & \cdots & g_{1N-6} & g_{1N-4} \\
h_{2N-3} & h_{2N-1} & \cdots & h_{2N-7} & h_{2N-5} & g_{1N-3} & g_{1N-1} & \cdots & g_{1N-7} & g_{1N-5} \\
\vdots & \vdots & \vdots & \vdots & \vdots & \vdots & \vdots & \vdots & \vdots & \vdots \\
h_0 & h_2 & \cdots & 0 & 0 & g_0 & g_1 & \cdots & 0 & 0 \\
0 & h_1 & \cdots & 0 & 0 & 0 & g_1 & \cdots & 0 & 0 \\
0 & h_0 & \cdots & 0 & 0 & 0 & g_0 & \cdots & 0 & 0 \\
\vdots & \vdots & \vdots & \vdots & \vdots & \vdots & \vdots & \vdots & \vdots & \vdots \\
0 & 0 & \cdots & h_{2N-2} & 0 & 0 & 0 & \cdots & g_{1N-2} & 0 \\
0 & 0 & \cdots & h_{2N-3} & h_{2N-1} & 0 & 0 & \cdots & g_{1N-3} & g_{1N-1}
\end{bmatrix}
$$

$$(12.4)$$

A requirement for the transform to be orthogonal (i.e., $\mathbf{V}^T\mathbf{V} = \mathbf{I}$) is that the sum of the squares of each column must be equal to one and the scalar product of different columns must be equal to zero. Therefore, for a filter bank that utilizes low-pass and high-pass filters, the following conditions ensure orthogonality of the transform so that no information is lost in the decomposition process (Strang and Nguyen, 1998):

$$
\sum_{N=0}^{2N-1-2l} h_n h_{n+2l} = \begin{cases} 1, & l = 0 \\ 0, & 0 < l < N \end{cases}
\tag{12.5a}
$$

$$
g_n = (-1)^{n+1} h_{2N-1-n}, n = 0, 1, \ldots, N-1
\tag{12.5b}
$$

with l an arbitrary point in the length of the sequence. If the filtering sequences satisfy these conditions, the entire structure is termed a quadrature-mirror filter (QMF) bank (Strang and Nguyen, 1998). A QMF bank is said to enjoy a perfect reconstruction (PR) property, because $\mathbf{x}$ can be reconstructed from $\mathbf{t}$, which means that there is no loss of information in the decomposition process. In fact, from the relation $\mathbf{t} = \mathbf{xV}$, it follows that $\mathbf{tV}^T = \mathbf{xVV}^T$ and $\mathbf{x} = \mathbf{tV}^T$, due to the orthogonality of the transform. Filtering in the wavelet transform consists of replacing some of the elements of $\mathbf{t}$ by zero so that a new vector $\bar{\mathbf{t}}$ is produced.

One limitation of the procedure described for the filtering of the time-domain signatures in the wavelet domain is that the wavelets must be chosen a priori and are not adapted to optimally describe the experimental data set. Optimizing the transform to maximize its compression ability and therefore its efficiency is normally achieved by optimizing the QMF bank (Daubechies, 1992; Vaidyanathan, 1993; Sherlock and Monro, 1998). A QMF bank is described by a set of parameters that can be adjusted by any algorithm for unconstrained optimization to maximize the compression ability of the transform. The parametrization of PR finite impulse response (FIR) filter banks proposed by Vaideanathan as adapted by Sherlock and Monro to parametrize orthonormal wavelets of arbitrary compact support may be used for the purpose. For the filter bank in Fig. 12.3 where the conditions in Eqs. (12.5a) and (12.5b) are satisfied, the transfer function of the low-pass filter in the z-transform domain can be written as:

$$H^{(N)}(z) = \sum_{n=0}^{2N-1} h_n^{(N)} z^{-n} = H_0^{(N)}(z^2) + Z^{-1} H_1^{(N)}(z^2) \tag{12.6}$$

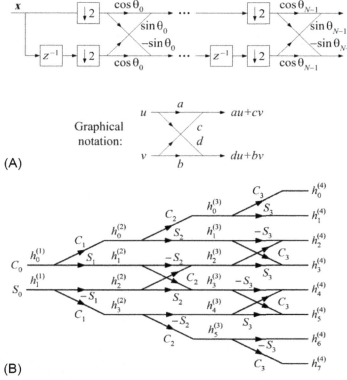

FIG. 12.3 (A) Procedure for parameterizing wavelet filter banks by N angles. (B) Recursive generation of low-pass filter weights $\{h_n^{(k+1)}\}$ in terms of $\{h_n^{(k)}\}$ by adding one additional angular parameter at a time. S_k and C_k represent the sine and cosine of angular parameter $h^{(k)}$, respectively. By using this algorithm, any set of N angles $\{\theta_0, \theta_1, \dots \theta_{N-1}\}$ leads to a sequence of low-pass filter weights that satisfies the orthogonality condition (12.5a). *(Reproduced from Froese, T., Hadjilou-cas, S., Galvao, R. K. H., Becerra, V. M., Coelho, C. J., 2006. Comparison of extrasystolic ECG signal classifiers using discrete wavelet transforms. Pattern Recogn. Lett. 27(5), 393–407.)*

where superscript (N) denotes that the filtering sequences have length $2N$. $H_0^{(N)}(z)$ and $H_1^{(N)}(z)$, termed polyphasic components of $H^{(N)}(z)$, are given by:

$$H_0^{(N)}(z) = \sum_{i=0}^{N-1} h_{2i}{}^{(N)} z^{-i} \tag{12.7a}$$

$$H_1^{(N)}(z) = \sum_{i=0}^{N-1} h_{2i+1}{}^{(N)} z^{-i} \tag{12.7b}$$

Defining the polyphasic components $G_0^{(N)}(z)$ and $G_1^{(N)}(z)$ of the high-pass filter $G^{(N)}(z)$ in a similar manner, a matrix $F^{(N)}(z)$ may be defined:

$$F^{(N)}(z) = \begin{bmatrix} H_0^{(N)}(z) & H_1^{(N)}(z) \\ G_0^{(N)}(z) & G_1^{(N)}(z) \end{bmatrix} \tag{12.8}$$

It can be shown (Sherlock and Monro, 1998; Tuqun and Vaidyanathan, 2000) that $F^{(N)}(z)$ can be factorized as:

$$F^{(N)}(z) = \begin{bmatrix} C_0 & S_0 \\ -S_0 & C_0 \end{bmatrix} \prod_{i=1}^{N-1} \begin{bmatrix} 1 & 0 \\ 0 & z^{-1} \end{bmatrix} \begin{bmatrix} C_i & S_i \\ -S_i & C_i \end{bmatrix}, \tag{12.9}$$

where each pair of parameters (C_i, S_i) are related to a common angular parameter θ_i as $C_i = \cos(\theta_i)$ and $S_i = \sin(\theta_i)$, $i = 0, 1, \ldots, N-1$. It follows that the filters can be completely parameterized by N angles $\theta_0, \theta_1, \ldots, \theta_{N-1}$, which can assume any value in the set of real numbers, as shown in Fig. 12.3B.

The weights of the low-pass filter can be easily recovered from a set of angles $\{\theta_i\}$ by using the following recursive formula (Vaidyanathan, 1993):

$$F^{(k+1)}(z) = F^{(k)}(z) \cdot \begin{bmatrix} 1 & 0 \\ 0 & z^{-1} \end{bmatrix} \begin{bmatrix} C_k & S_k \\ -S_k & C_k \end{bmatrix} \tag{12.10}$$

for $k = 1, 2, \ldots, N-1$ with.

$$F^{(1)}(z) = \begin{bmatrix} C_0 & S_0 \\ -S_0 & C_0 \end{bmatrix}. \tag{12.11}$$

Eq. (12.10) with the initial condition of Eq. (12.11) provides a way to obtain the weights $\{h_i^{(k+1)}\}$ for a filter of length $2(k+1)$ from the weights $\{h_i^{(k)}\}$ for a filter of length $2k$. Using Eqs. (12.8) and (12.10), it follows that

$$H_0^{(k+1)}(z) = H_0^{(k)}(z)C_k - z^{-1}H_1^{(k)}(z)S_k, k = 1, 2, \ldots, N-1 \tag{12.12a}$$

$$H_1^{(k+1)}(z) = H_0^{(k)}(z)S_k - z^{-1}H_1^{(k)}(z)C_k, k = 1, 2, \ldots, N-1, \tag{12.12b}$$

where $H_0^{(1)}(z) = C_0$ and $H_1^{(1)}(z) = S_0$. A recursive formula for the generation of low-pass filter weights with even indexes $\{h_{2i}\}$ can be stated by using the definitions in Eqs. (12.7a) and (12.7b) to expand Eq. (12.12a):

$$\overbrace{\sum_{i=0}^{k} h_{2i}^{(k+1)} z^{-i}}^{H_0^{(k+1)}(z)} = \overbrace{\left(\sum_{i=0}^{(k-1)} h_{2i}^{(k)} z^{-1} \right)}^{H_0^{(k)}(z)} C_k - Z^{-1} \overbrace{\left(\sum_{i=0}^{(k-1)} h_{2i+1}^{(k)} z^{-1} \right)}^{H_1^{(k)}(z)} S_k \Rightarrow$$

$$\sum_{i=0}^{k} h_{2i}^{(k+1)} z^{-i} = C_k h_0^{(k)} + \sum_{i=1}^{k-1} \left(C_k h_{2i}^{(k)} - S_k h_{2i-1}^{(k)} \right) z^{-i} - S_k h_{2k-1}^{(k)} z^{-k}$$

$$(12.13)$$

for $k = 1, 2, ..., N-1$, with $h_0^{(1)} = C_0$ and $h_1^{(1)} = S_0$. From the identity of terms with the same power of z in the last line of Eq. (12.13), it follows that:

$$\begin{cases} h_0^{(k+1)} &= C_k h_0^{(k)} \\ h_{2i}^{(k+1)} &= C_k h_{2i}^{(k)} - S_k h_{2i-1}^{(k)}, \quad i = 1, 2, ..., k-1 \\ h_{2k}^{(k+1)} &= S_k h_{2k-1}^{(k)} \end{cases} \quad (12.14a)$$

for $k = 1, 2, ..., N-1$. A similar formula can be stated for the low-pass filter weights with odd indexes, by expanding Eq. (12.12b) as.

$$\begin{cases} h_1^{(k+1)} &= S_k h_0^{(k)} \\ h_{2i+1}^{(k+1)} &= S_k h_{2i}^{(k)} - C_k h_{2i-1}^{(k)}, \quad i = 1, 2, ..., k-1 \\ h_{2k+1}^{(k+1)} &= C_k h_{2k-1}^{(k)} \end{cases} \quad (12.14b)$$

for $k = 1, 2, ..., N-1$. After obtaining the low-pass filtering sequence as explained above, the high-pass filtering sequence can be obtained in a similar manner using Eq. (12.5b). It is worth noting that from Eq. (12.2), it can be seen that the highpass filter coefficients can also be directly obtained from the lowpass ones.

In addition, because of the orthogonality of the transform from Eq. (12.1) one can observe that the $2N$ weights $\{h_n\}$ of the highpass filter are subject to N restrictions. Thus, there are N degrees of freedom that can be used to optimize the filter bank according to some performance criterion. Since the restrictions are nonlinear and may define a nonconvex search space, the optimization task is not trivial. We circumvent this difficulty by using the lattice structure for the filter bank, which is parameterized by N angles $\{\theta_0, \theta_1, ..., \theta_{N-1}\}$ that can assume any real value as shown in Fig. 12.3.

The problem then becomes one of unconstrained optimization in R^N. The optimal filtering procedure employed in this example is aimed at maximizing the amount of energy in the wavelet coefficients kept in the thresholding process. The optimization consisted of maximizing an objective function $F(\theta): R^N \to R$ defined as.

$$F(\theta) = \sum_{k \in I} p^2(k; \theta) \quad (12.15)$$

where $\boldsymbol{\theta}$ is the vector of N angles that parameterize the filter bank as explained above, $p(k;\boldsymbol{\theta})$ is the kth wavelet coefficient resulting from the signal decomposition, and I is the index set of the coefficients kept in the thresholding process. It is worth noting that I is defined with basis on the magnitude of the wavelet coefficients before the optimization. The flexible polyhedron algorithm available in the MATLAB Optimization Toolbox may be employed to search for the optimum $\boldsymbol{\theta}$, using the parameters associated with the db4 wavelet as a starting point.

The algorithm has been sucessfully employed within a chemometrics context in the work by Galvão et al. (2004) to explain X and Y data statistics as well as for the optimal discrimination and classification of THz spectra in the wavelet domain in continuous-wave THz spectrometry (Galvão et al., 2003) and in the work by Froese et al. (2006) for the classification of electrocardiography (ECG) signals and for electromyographic signals (Paiva et al., 2008). In addition, it has been used in system identification of broadband pulse propagation in waveguides in experiments performed using femtosecond pulse spectrometry (Hadjiloucas et al., 2004). More recent developments in the theory of orthonormal filter banks include the use of additional constraints to ensure three vanishing moments (Paiva et al., 2009; Uzinski et al., 2013). The algorithm was also sucessfully employed in a chemometrics context to comress ECG signals to imporve classification (Hadjiloucas et al., 2014).

12.2.2 Use of wavelet transforms for system identification based chemometrics in time-domain spectrometry

One of the more interesting developments of the wavelet transform is in its application to dynamic systems when these are described in state space form (Paiva and Galvão, 2006; Uzinski et al., 2017). These algorithms can be used to place chemometrics for time-domain spectroscopy in a systems identification framework (Yin et al., 2007, 2008; Hadjiloucas et al., 2009). Defining the background and sample interferograms as the input and output signals, the frequency response of an identified model would be an estimate of the complex insertion loss (CIL). A wavelet-packet formulation illustrated in Fig. 12.4 is adopted and sub-band models $M_{i,j}(z)$ are identified from the sample and background interferograms by following a least-squares procedure as indicated in Fig. 12.5.

Fig. 12.5 illustrates the procedure adopted to identify each sub-band model $M_{i\ j}$. $\mathbf{u}$ is the input signal used for identification; $\mathbf{y}$ and $\breve{u}_{i,j}$ are the plant and sub-band model outputs, respectively. Residue $e_{i\ j}$ denotes the wavelet-packet coefficients of the difference between $\mathbf{y}$ and $\breve{u}_{i,j}$, in the frequency band under consideration. The structure adopted for the sub-band model is a transfer function of the form:

$$M_{i,j}(z) = P_{i,j}(z)Q_{i,j}(z), \tag{12.16}$$

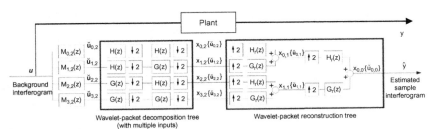

FIG. 12.4 Wavelet-packet model structure. In this example, a complete two-level decomposition tree, which defines four frequency sub-bands, is employed. $H(z)$, $G(z)$ denote low-pass and high-pass decomposition filters, respectively, with reconstruction counterparts represented by $H_r(z)$, $G_r(z)$. The four sub-band models are represented by the transfer functions $M_{0,2}(z)$, $M_{1,2}(z)$, $M_{2,2}(z)$, $M_{3,2}(z)$. *(Reproduced from Paiva H. M., Galvão R. K. H., 2006. Wavelet-packet identification of dynamic systems in frequency subbands, Signal Process. 86, 2001–2008.)*

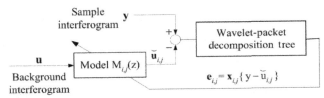

FIG. 12.5 Model identification of a sample interferogram for a given frequency sub-band. *(Reproduced from Paiva H. M., Galvão R. K. H., 2006. Wavelet-packet identification of dynamic systems in frequency subbands, Signal Process. 86, 2001–2008.)*

where:

$$P_{i,j}(z) = \left(\frac{1}{1-z^{-1}} \right)^{s_{i,j}}, \, s_{i,j} \in Z \qquad (12.17a)$$

$$Q_{i,j}(z) = \alpha_{ij} + \beta_{ij}z^{-1}, \, \alpha_{i,j}, \beta_{ij} \in \Re \qquad (12.17b)$$

$P_{i,j}(z)$ is aimed at roughly approximating the band-limited frequency response of the plant, whereas the Finite Impulse Response (FIR) term $Q_{i,j}(z)$ provides a fine-tuning for the approximation. A least-squares adjustment for the parameters of $M_{i,j}$ can be carried out by minimizing the following cost function $J_{i,j}$: $Z \times \Re^2 \rightarrow \Re$:

$$J_{i,j}\left(s_{i,j}, \alpha_{i,j}, \beta_{i,j}\right) = e_{i,j}\left(e_{i,j}\right)^{\mathrm{T}}, \qquad (12.18)$$

where $e_{i,j}$ denotes the row vector of residues for the identification data, as shown in Fig. 12.5. As discussed elsewhere (Paiva and Galvão, 2006), if $S_{i,j}$ is fixed, the optimal real-valued parameters $\alpha_{i,j}^*$ and $\beta_{i,j}^*$ are obtained by imposing:

$$\frac{\partial J_{i,j}}{\partial \alpha_{i,j}} = \frac{\partial J_{i,j}}{\partial \beta_{i,j}} = 0 \qquad (12.19)$$

To find the optimal value of $S_{i,j}$ the following search algorithm in Z is used: the value of $S_{i,j}$ is varied in a specified range. For each value of $S_{i,j}$, the optimal values $\alpha_{i,j}^{*}$ and $\beta_{i,j}^{*}$ are calculated using.

$$\begin{bmatrix} \alpha_{i,j}^{*} \\ \beta_{i,j}^{*} \end{bmatrix} = M^{-1} \begin{bmatrix} x_{i,j}\{y\}\left(x_{i,j}\left\{u_{i,j}^{p}\right\}\right)^{T} \\ x_{i,j}\{y\}\left(x_{i,j}\left\{u_{i,j}^{pd}\right\}\right)^{T} \end{bmatrix} \tag{12.20}$$

with

$$M = \begin{bmatrix} x_{i,j}\left\{u_{i,j}^{p}\right\}\left(x_{i,j}\left\{u_{i,j}^{p}\right\}\right)^{T} & x_{i,j}\left\{u_{i,j}^{pd}\right\}\left(x_{i,j}\left\{u_{i,j}^{p}\right\}\right)^{T} \\ x_{i,j}\left\{u_{i,j}^{p}\right\}\left(x_{i,j}\left\{u_{i,j}^{pd}\right\}\right)^{T} & x_{i,j}\left\{u_{i,j}^{pd}\right\}\left(x\left\{u_{i,j}^{pd}\right\}\right)^{T} \end{bmatrix} \tag{12.21}$$

provided M^{-1} exists. In these equations, $u_{i,j}^{pd}$ denotes the value of $u_{i,j}^{p}$, which is the output term of $P_{i,j(z)}$ delayed by one sample. The value $s_{i,j}$ for which $J_{i,j}$ is minimum is then adopted, as well as the corresponding values of $\alpha_{i,j}^{*}$ and $\beta_{i,j}^{*}$.

The structure of the wavelet decomposition tree can be optimized in order to achieve a compromise between the parsimony and accuracy of the overall model. For this purpose, a generalized crossvalidation procedure can be employed to determine whether the reduction in identification error is large enough to justify the further decomposition of any given tree node (Paiva and Galvão, 2006). Fig. 12.6 shows results of the calculated complex insertion loss function from a THz transient spectrometer using (a) standard FFT-based procedures, (b) subspace identification, and (c) the wavelet-packet identification procedure (Yin et al., 2008). It is worth noting that the tree structure was automatically defined by the identification algorithm, with no prior knowledge of the spectral features of the sample under consideration. A dispersion metric and a discrimination metric can be subsequently used to discriminate between samples on the basis of the identified wavelength coefficients (Yin et al., 2008). According to these metrics, the identification methods are seen to be more robust with respect to noise than the standard ratioing procedure. Furthermore, the proposed wavelet-packet identification technique becomes slightly superior to the subspace method at larger noise levels.

The wavelet packet identification work is also of considerable relevance to the previous chapters in the book as it provides a methodology, which can be directly adopted for Kalman filtering. The Kalman filter community has already used wavelet transforms in several applications, e.g., for load forecasting in electric power systems (Zheng et al., 2000; Caujolle et al., 2010), for short-term traffic volume forecasting (Xie et al., 2007), for GPS cycle slip correction (Collin and Warnant, 1995), for panchromatic sharpening of remote-sensing images (Garzelli and Nencini, 2007), for real-time flood forecasting (Chou and Wang, 2004), or for respiration rate estimation (Johnson et al., 2013).

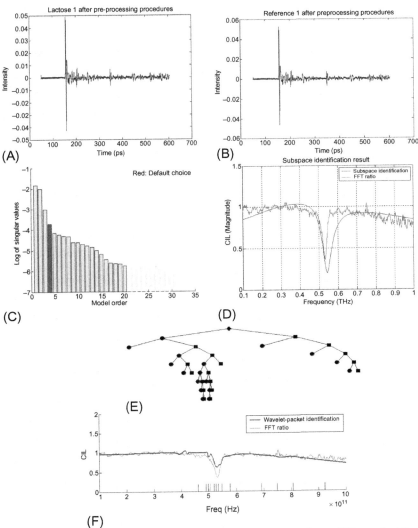

FIG. 12.6 (A) Reference (background) time domain interferogram from a THz transient time-domain spectrometer, (B) sample interferogram, (C) singular value plot for the subspace identification and (D) calculated magnitude of the complex insertion loss for the lactose sample as a function of frequency obtained by ratioing the sample spectrum against the background spectrum *(lighter grey line)* and by subspace identification *(darker smoother line)*, (E) resulting wavelet-packet tree and (F) CIL for a lactose sample obtained by wavelet-packet identification *(darker smoother line)*. The FFT ratio result *(lighter grey line)* is also presented for comparison. The frequency-domain segmentation automatically defined in the identification procedure is indicated by vertical lines at the bottom of the graph. As can be seen, the segmentation is more refined in the spectral region corresponding to the absorption band. *(Results reproduced from Yin, X.X., Hadjiloucas, S., Fischer, B.M., Ng, B.W.H., Paiva, H.M., Galvao, R.K.H., Walker, G.C., Bowen, J.W., Abbott, D., 2008. Classification of lactose and mandelic acid THz spectra using subspace and wavelet-packet algorithms, Proceedings of the SPIE 6798, Microelectronics: Design, Technology, and Packaging III, 679814.)*

Furthermore, in many respects, the function of Kalman filtering can be comparable to wavelet filtering for denoising (Tangborn and Zhang, 2000; Postalcloglu et al., 2005; Renaud et al., 2005; Salis et al., 2013) and both may be conveniently implemented equally well using reconfigurable systolic arrays (Sudarsanam et al., 2010). A wavelet ensemble Kalman filter has also been proposed for data assimilation in spatial models that uses diagonal approximation of the state covariance in the wavelet space to achieve adaptive localization (Beezley et al., 2011). Finally, the asymptotic convergence of adaptive learning rates of extended Kalman filter-based training algorithms of wavelet neural networks may be ensured using algorithms that combine both concepts (Kim et al., 2006).

12.2.3 Chemometrics in the context of optimizing integration time for resolving spectral features according to tolerance specifications

In the following section, we discuss a way of using wavelet transforms to establish a criterion for establishing the integration time or number of coaveraged interferograms in Fourier transform spectroscopy on the basis of a olerance error level in the measurement process (Galvão et al., 2002). Use of the statistical properties of the wavelet-transform coefficients for optimization of integration time in Fourier transform spectrometry. In this example, a non-optimized filtering procedure, where each time-domain signal x to be filtered was decomposed using a db4 wavelet filter bank with two resolution levels, was adopted. Hard thresholding was used, with all wavelet coefficients with magnitude smaller than a certain fraction of the largest coefficient being replaced with zeros. The inverse transform was then applied to obtain the filtered signal.

Choosing the threshold is a nontrivial task, since removing too many coefficients leads to distortion in the filtered signal. However, a convenient threshold value can be obtained by simultaneously analyzing the acquired time domain interferograms during the coaveraging process in order to discard wavelet coefficients with absolute value not significantly larger than zero from a statistical point of view. Suppose that k interferograms have been acquired up to the present moment so that the coaveraged signal is $\bar{I}[k]$. The wavelet transform is applied to the k interferograms and the mean $\mu_{t_i}[k]$ and total standard deviation $\sigma_{t_i}[k] = s_{t_i}[k]/\sqrt{k}$ of each wavelet coefficient t_i $(i = 1, 2, \ldots N)$ are evaluated, where $s_{t_i}[k]$ is the standard deviation of t_i over the k interferograms. The notation implies that the double index (scale, position) of the wavelet coefficients was collapsed into a single index i. Only those coefficients for which $|\mu_{t_i}[k]| > 2.326\sigma_{t_i}[k]$ are retained (99% confidence level for the hypothesis test $|t_i| > 0$, since a cumulative normal distribution function $P(x)$ equals 0.99 for $x = 2.326$ standard deviations). During the coaveraging process, the overall signal-to-noise ratio in the wavelet coefficients SNR_t for the filtered case can also be estimated as:

$$SNR'_t[k] = \frac{\sqrt{\dfrac{1}{N}\displaystyle\sum_{i=1}^{N}\mu_{t_i}^{2}[k]H\{|\mu_{t_i}[k]| - 2.326\sigma_{t_i}[k]\}}}{\sqrt{\dfrac{1}{N}\displaystyle\sum_{i=1}^{N}\sigma_{t_i}^{2}[k]H\{|\mu_{t_i}[k]| - 2.326\sigma_{t_i}[k]\}}},$$ (12.22)

where $H(x)=0$, $x<0$ and $H(x)=1$, $x>0$ (Heaviside unit step function). For the unfiltered case, all wavelet coefficients are used for the calculation of SNR_t' (the Heaviside function is not used). If $SNR_t[k]$ is larger than a tolerance level, the data acquisition process is stopped whereas if it is smaller, the process is repeated for $SNR_t'[k+1]$. As an example, a coaveraging process was simulated by adding white Gaussian noise to the time domain signature of a poly-vinyl-chloride (PVC) sample as recorded using a continuous wave terahertz dispersive Fourier transform spectrometer. The top interferogram shown in the inset of Fig. 12.7 was recorded after coaveraging 200 waveforms, whereas the degraded one simulates a typical single scan. A clean background interferogram with an SNR of 800 was also recorded (this is typical for a well-aligned system employing a mercury arc lamp source and a helium-cooled hot electron bolometer detector). The wavelet transform was performed in MATLAB using the "db4" mother wavelet with 2 scale levels.

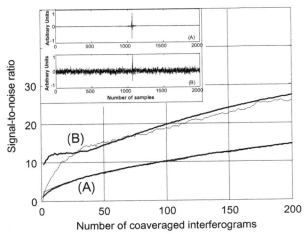

FIG. 12.7 Signal-to-noise ratio for the coaveraged interferogram (A) without and (B) with wavelet filtering. The thick lines are estimates SNR_t' and the thin lines are the true values SNR_t. The inset shows (A) the sample time-domain signal and (B) an interferogram degraded by reducing the signal-to-noise ratio by a factor of 50. (*Results reproduced from Galvão R.K.H., Hadjiloucas S., Bowen J.W., 2002. Use of the statistical properties of the wavelet transform coefficients for the optimisation of integration time in Fourier transform spectrometry, Opt. Lett. 27, 643–645.*)

The main graph in Fig. 12.7 displays the true SNR_t and estimated SNR_t' both before and after wavelet filtering. The true SNR_t can be calculated since, in this simulated example, the signal and the artificial noise added are known. After 20 interferograms have been coaveraged, the statistical fluctuations settle down and the estimates SNR_t' become quite close to the true values. Thus, the estimation procedure can be used as a reliable guideline to stop the data acquisition process. In addition, the filtering process increases both SNR_t and SNR_t', thus allowing a smaller number of coaveragings to be performed for a given signal quality criterion. As shown in Fig. 12.8, there is an improvement in signal quality by filtering, since the sum-squared difference between the coaveraged spectrum and the true spectrum decreases after filtering. However, after a large number of coaveragings (crossing point in the graph), the filtering process may cease to be advantageous, since the noise level is already so low that further removal of noise does not compensate for the small distortions caused by the filtering process. The inset in Fig. 12.8 is a quadratic fit to the gradual change of the crossing point when different RMS noise levels are added to individual interferograms before starting the coaveraging process (this corresponds to an observed shift toward the right in the main graph).

Once calibration spectra have been acquired, a prediction model can be built by linear regression to relate the complex insertion loss $\hat{L}(\lambda)$ at M spectral bins with center wavelength λ to the physical parameter of interest y (the chemical

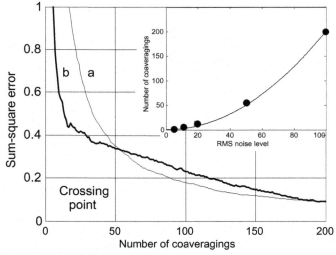

FIG. 12.8   Sum-square error of the coaveraged (thin line) and filtered (thick line) interferograms with respect to the original sample interferogram. The inset shows the gradual change of the crossing point with the increase in the RMS noise level (with respect to the noise level in the interferogram shown in the inset (A) of Fig. 12.7) added to individual interferograms during the coaveraging process. *(Results reproduced from Galvão R.K.H., Hadjiloucas S., Bowen J.W., 2002. Use of the statistical properties of the wavelet transform coefficients for the optimisation of integration time in Fourier transform spectrometry, Opt. Lett. 27, 643–645.)*

composition of the sample, for instance). If further enhancement in the signal-to-noise ratio of the complex insertion loss is required, box-car or Savitsky-Golay filtering may be implemented, at the cost of distortion in sharp features of the spectrum. Alternatively, to remove the residual noise and also to alleviate collinearity problems, which tend to increase the propagation of instrumental noise, it has recently been shown (Alsberg et al., 1997, 1998) that there are advantages in identifying the model in the wavelet domain. In this case, the regression model takes the form $y' = \sum_{j=1}^{n} w_j t_j$, where y' is an estimate of y, w_j is a weight parameter (the slope in the regression), and $\{t_j, j = 1, \ldots, n\}$, for $n < M$ is a subset of wavelet coefficients resulting from the decomposition of $\hat{L}$. The use of a reduced number of coefficients minimizes the risk of overfitting. This subset is chosen according to a selection criterion (Alsberg et al., 1997), which takes into account not only the statistics of the wavelet coefficients, but also correlations with y.

After the model has been obtained, it is possible to devise a stopping criterion for the coaveraging process based on the uncertainty bounds for the model predictions. This specification is more precise than the standard *SNR* objective that is commonly used for the determination of the number of interferograms that need to be used in the coaveraging process. To do that, one can assume that the calculated insertion loss $L(\lambda)$ has a noise $\eta(\lambda)$ so that its true value fluctuates around a mean value $\bar{L}$, i.e., $L = \bar{L} + \eta$. Suppose that this noise is white Gaussian and stationary with mean $E[\eta(\lambda)] = 0$ and variance $E[\eta^2(\lambda)] = \upsilon$ and that noise at different wavelengths is uncorrelated, that is, $E[\eta(\lambda_1)\eta(\lambda_2)] = 0 \, \forall \, \lambda_1 \neq \lambda_2$. Due to the orthogonality of the wavelet transform, the noise will also be uncorrelated across the wavelet coefficients and thus it follows that the predicted value of y will have a variance $\upsilon_{y'}$ given by:

$$\upsilon_{y'} = E\left[(y' - \bar{y'})^2 \right] = E\left[\left(\sum_{j=1}^{n} w_j \eta_{t_j} \right)^2 \right] = \sum_{j=1}^{n} w_j^2 \upsilon_{t_j}, \tag{12.23}$$

where η_{t_j} (with variance υ_{t_j}) is the noise on the jth wavelet coefficient. Estimated values υ_{t_j}' for υ_{t_j} can be obtained from the statistical properties of the respective wavelet transform coefficients during co-averaging. Data acquisition can be stopped when $\sqrt{\sum w_j^2 \upsilon_{t_j}'[k]} < Tol|\bar{y}[k]|$, where $\bar{y}[k]$ is the model prediction calculated after coaveraging k interferograms processed to give $L_\lambda [k]$ and $0 < Tol < 1$ is a specified tolerance level. If this tolerance is not met, another interferogram is acquired and the process is repeated. In order to apply this stopping criterion, each time a new time-domain interferogram $I[k]$ is acquired, this has to be fully processed until the wavelet coefficients of the insertion loss $L_\lambda [k]$ are calculated. The coefficients $t_j[k]$ of the wavelet transform of $L_\lambda [k]$ are calculated and then used to update the estimate υ_{t_j}'. Since only simple mathematical operations are involved, this whole procedure can be carried out within the time frame of a single interferogram acquisition. This implies that

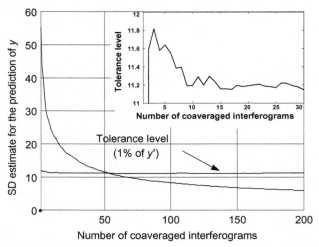

FIG. 12.9   Standard deviation SD estimate for the prediction of y. This was obtained by a coaveraging process after degrading the *SNR* of the original interferogram by a factor of 20 and setting all model weights w_j to one. The tolerance level, magnified in the inset, was obtained as 1% of the model prediction at each coaveraging step. *(Results reproduced from Galvão R.K.H., Hadjiloucas S., Bowen J.W., 2002. Use of the statistical properties of the wavelet transform coefficients for the optimisation of integration time in Fourier transform spectrometry, Opt. Lett. 27, 643–645.)*

the algorithm is well suited to imaging applications where real-time processing is important.

The standard deviation SD estimate for the prediction of y as a function of number of coaveraged interferograms is shown in Fig. 12.9. The inset shows that the tolerated 1% $|\bar{y}'[k]|$ converges after a certain number of coaveragings. Thus, it is possible to extrapolate the curve of model prediction variance estimate in order to predict the number of coaveraged interferograms required to reach the specified tolerance level and thus the time required for the measurements. Extrapolation is straightforward since the standard deviation of the noise in the coaveraged spectrum, and thus in the wavelet coefficients, decreases with the square root of the number of coaveraged spectra. A more comprehensive discussion of wavelet denoising of coaveraged spectra employing statistics from individual scans can be found in the article by Galvão et al. (2007b).

The chemometric approach mentioned here may be further extended using the more recent developments in adaptive wavelets stated in the previous section. Furthermore, it can be adopted for fast pulse time domain spectrometry with femtosecond pulses, where the wavelets can be used to perform apodization. Apodization functions are important in time domain spectrometry as they enhance the resolution of the spectra. The design of these window functions is a tradeoff of minimizing sidelobes of the calculated spectrum, resulting from the leakage of energy across different frequency bins, at the expense of increasing main lobe width, which leads to a reduction in frequency resolution. A quadratic programming optimization procedure for designing asymmetric apodization windows superior to conventional rectangular, triangular (Mertz), and Hamming windows, and tailored to the shape of time-domain sample waveforms is reported in Galvão et al. (2007a). The approach proposed therein uses Gaussian functions as the building elements of the apodization window (Fig. 12.10), but the formulation is sufficiently generic to accommodate other basis functions such as wavelets. A Kalman filter-based approach on the basis of

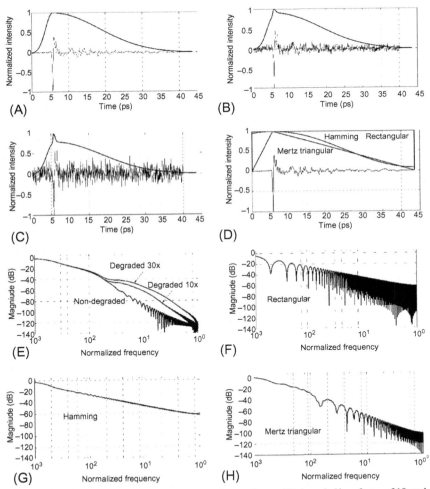

FIG. 12.10   (A) Nondegraded time-domain sample waveforms, (B) degraded by a factor of 10, and (C) degraded by a factor of 30 with corresponding optimized apodization windows. (D) Nondegraded waveform with conventional apodization windows. (E), (F), (G), (H) Corresponding frequency domain plots (low-frequency value normalized to 0 dB). *(Reproduced from Galvão, R. K. H., Hadjiloucas, S., Zafiropoulos, A., Walker, G. C., Bowen, J. W., Dudley, R., 2007a. Optimization of apodization functions in terahertz transient spectrometry. Opt. Lett. 32(20), 3008–3010.)*

quadratic programming for these window functions may also be developed to improve the resolution of interferograms accumulated in fast scan mode.

12.3 Recent advances in classification and clustering

12.3.1 Chemometrics approaches using orthogonal or oblique projections: PCA and ICA

Classification and clustering methods are particularly relevant to chemical pattern recognition. However, not all problems of interest to the analyst are about classifying groups of samples. For example, there may be cases where spatial variability regarding the location of collection of samples might have to be taken into account. The chemical composition of these samples might be influenced by various sources of contamination. Furthermore, it could also be the case that the analyst does not wish to classify samples into contaminated and noncontaminated, but rather to determine the main factors that influence the composition of the samples (Brereton, 1987).

Factor analysis (Malinowski and Howery, 1980; Horst, 1965; Rummel, 1970) is also an important area of chemometrics. Within the context of supervised classification methods, principal components is used largely to reduce the dimensionality of the data. However, often the factors in themselves may be of interest. For example, the variability between samples may be due to a multitude of factors. Factor analysis is essentially also a principal component-based method. Firstly, PCA is performed on the data set, appropriately preprocessed and scaled. The next step is to determine the significant number of factors. These factors may correspond to physically meaningful processes or components. A typical example is mixture analysis where each factor ideally corresponds to one component in a mixture. Most factor analysis algorithms rank the factors according to their contribution to the overall variability of the data set. A simple example is of a peak arising from a single compound (Brereton, 1987).

Although multiple measurements of intensity across the peak might be made, apart from experimental noise these measurements will maintain exactly the same proportions in each sample if the peak corresponds to one pure component. Hence, in the absence of noise, all measurements across the peak should be reduced to one factor corresponding to 100% of the variability. In practice, however, there will be some noise reflected in random fluctuations in peak intensity, so the actual picture will not be so clear. The first factor might correspond to 90% of the variability, the second 5%, and so on. So the question posed by the analysis is whether we can unambiguously state that there is only one factor. But from a chemometrics perspective, there are a large number of criteria for estimating the number of significant factors. A very simple test is a 95% cutoff, i.e., all factors are considered significant until their sum is greater than 95% of the total variability. This method does not allow for differing noise levels and distributions, and a more satisfactory method has been proposed by Malinowski (1987). Several other criteria have also been proposed, but the best method depends on prior knowledge of noise and signal distributions in a similar way to optimum filters as discussed in the signal processing literature.

A completely different criterion is crossvalidation (Wold, 1978), and this probably works best when the noise is hard to model. Some analogy between filters versus maximum entropy and best numerical criteria for estimating the number of significant factors and crossvalidation can be made. An example where several methods are compared for the estimation of significant number of factors in mass spectrometry and nuclear magnetic resonance spectroscopy of mixtures can be found in the work by Hearmon et al. (1987), and references therein.

Once the number of significant factors has been estimated it is often desired to interpret these factors in terms of physical processes. For example, in mixture analysis each factor should correspond to the spectrum of each component of the mixture. The simplest way to do this is by rotation, where PCA is used to determine the dimensionality of the space so that the principal components are then rotated in the PCA space. Orthogonal rotation preserves the geometry of the space, i.e., in practical terms scaling of the raw data is preserved. There are various criteria for rotation, but the main objective is to try to maximize the variability along each factor, and the clustering of objects around a given new principal component. There are two principal types of orthogonal rotation, namely, varimax (Kaiser, 1958) and quartimax (Harman, 1967) differing according to the definition of variability.

More flexible methods involve oblique rotations where the geometry is not necessarily preserved. These include quartimin, (Carroll, 1957) oblimax (Saunders, 1960), and covarimin (Kaiser, 1958). Successive rotations may also be performed to the dataset (Hadjiloucas et al., 2002). As discussed in the article by Daszykowski (2007), these techniques belong to a wider family of projection algorithms (PA) and enable similar latent factors (PCs) to be constructed. In general, the latent factors can be viewed as a special solution of the projection pursuit approach (PP) (Friedman and Tuckey, 1974). The main goal of PP is to find a set of the low-dimensional projections (latent factors) that maximize the so-called projection index (PI), which defines the intent of the method. The search for directions that maximize the PI of a given projection can be facilitated, using the PA (Croux and Ruiz-Gazen, 1996).

$$\mathbf{X} = \sum_{i=1}^{f} \mathbf{t}_i \mathbf{p}_i^T + \mathbf{E}, \tag{12.24}$$

where the data, $\mathbf{X}$, are represented as a product of two matrices, with the columns of the first matrix being the scores, $\mathbf{t}_i$, and the columns of the second (loading) matrix are the weight vectors, $\mathbf{p}_i$, which explain contributions of the individual data variables to construction of each latent factor. The residual matrix, **E**, describes the part of the data, which remains unexplained by the model with f latent factors. This is done in such way that every latent factor is the weighted sum of the explanatory variables (in other words, it is their linear combination). From a geometrical point of view, the loading vector, being of unit length, points out to the direction in the multivariate data space, whereas the score vector is a result of an orthogonal data projection onto that direction (Glover and Hopke, 1992). An interesting more recent development in multicomponent analysis is the successive projections algorithm discussed by Araújo et al. (2001), which more recently was also used for classification of resistive, capacitive as well as inductive networks (Galvão et al., 2013a, b; Galvão et al., 2018; Jacyntho et al., 2015). These networks may be used to emulate dielectric mixture responses in a quality control context (Galvão et al., 2013a, b; Galvão et al., 2017).

Nowadays, independent component analysis algorithms (ICA), which are nonorthogonal but instead oblique projections (Cichocki and Amari, 2002) can also be adopted for chemometrics (Hyvärinen et al., 2001). In ICA, the observed data are assumed to be generated by mixing a number of independent sources Hyvärinen and Oja, 2000, De Lathauwer et al., 2000). The main goal of this method is to provide a set of latent variables, which are constructed to be statistically independent. Statistical independence is a stronger condition than orthogonality. Two variables can be orthogonal, but still statistically dependent. In order to determine a set of statistically independent components, a PI maximizing a certain measure of non-Gaussicity is used. The most popular one is entropy, which makes a direct link between the PP and the ICA approach. The entropy of a projection can be approximated, using the higher-order moments of the data distribution:

$$PI(\mathbf{t}_i) = \frac{1}{12}E_3\{\mathbf{t}_i^2\} + \frac{1}{48}\text{kurtosis}(\mathbf{t}_i)^2, \tag{12.25}$$

where E_3 is the third moment of the data distribution. The ICA approach is mostly applied in the field of signal processing, but it starts gaining popularity in chemistry, basically for identification of pure components in mixtures (Visser and Lee, 2004; Westad and Kermit, 2003; Wang et al., 2006; Westad, 2005; Chen and Wang, 2001).

A particularly new approach that links state estimation as discussed in the previous chapters with independent component analysis can be found in the article by Odiowei and Cao (2010) as applied to the Tennessee Eastman Process plant (Chen et al., 2004). Furthermore, ICA has been discussed within the context of state space feedforward and feedback structures for blind source recovery (Salam and Waheed, 2001; Waheed and Salam, 2003). In many respects these algorithms are likely to play quite a fundamental role in the future development of Chemometrics within the context of state estimation.

12.3.2 Advances in support vector machine classifiers

From a clustering analysis perspective, much progress has also taken place recently using supervised learning techniques (Vapnik, 1998, 1999) such as support vector machine (SVMs). These algorithms perform a mapping of lower-dimensional datasets into a high-dimensional feature space where a separating hyperplane, which maximizes the boundary margin between two classes, can be established (Yin et al., 2017; Burges, 1998; Joachims, 1998). Although originally SVMs have been used for binary classification tasks, recent extensions discuss their use to multiclass classification problems (Weston and Watkins, 1999; Platt et al., 2000; Fei and Liu, 2006).

In their simplest form, as binary classifiers, SVMs classify data on the basis of a set of support vectors. The support vectors are subsets of the training data sets and are used to construct the afforementioned hyperplane in feature space, which acts as a boundary separating the different classes.

Linear SVM classification is often performed by using a real-valued function $f : X \subseteq \mathbb{R}^n \to \mathbb{R}^n$, where the input $X = (x_1, \ldots, x_n)'$ is assigned to a positive class if $f(x) \geq 0$, and to a negative class if $f(x) \leq 0$. The $f(x)$ function is a linear mapping of $x \in X$, which can be written as:

$$f(x) = \text{sgn}\langle w, x \rangle + b = \sum_{i=1}^{n} w_i x_i + b \tag{12.26}$$

where $(w, b) \in \mathbb{R}^n \times \mathbb{R}$ are the parameters that control the function $f(x)$, x_i refers to examples (training samples), and b is the bias term of the hyperplane used to separate binary datasets (Platt et al., 2000).

The optimal separating hyperplane (OSH), can be obtained by maximizing a margin:

$$\min_{w, b} \frac{1}{2} \|w\|^2 \tag{12.27a}$$

subject to:

$$y_i(w^* \cdot x_i + b) - 1 \geq 0, \quad i = 1, \ldots, m \tag{12.27b}$$

From Eq. (12.26), it can be found that $y_i(w \cdot x_i + b) \geq 0$ for $y_i = +1$, while $y_i(w \cdot x_i + b) < 0$ for $y_i = -1$, where y_i is the class label, and w the weight vector.

Eq. (12.27a) can be given in the form of a Lagrange function:

$$L(w, b, \alpha) = \frac{1}{2} \|w\|^2 - \sum_{j=1}^{N} \alpha_i [y_i(w \cdot x_i + b) - 1], \tag{12.28}$$

where α_i are the Lagrange multipliers. The goal here is to minimize the Lagrange function (12.28) with respect to w and b, and maximize (12.30) with respect to $\alpha \geq 0$.

The minimum with respect to w and b of the Lagrangian in Eq. (12.28) is given by imposing the following two conditions:

$$\frac{\partial L}{\partial w} = 0 \Rightarrow w = \sum_{i=1}^{N} \alpha_i y_i x_i \tag{12.29a}$$

and

$$\frac{\partial L}{\partial b} = 0 \Rightarrow w = \sum_{i=1}^{N} \alpha_i y_i = 0 \tag{12.29b}$$

The following equation shows the condition satisfying both constraints in Eqs. (12.29a) and (12.29b); this is also referred within the SVM classifier literature as the dual problem:

$$\max_{\alpha} L(\alpha) = \max_{\alpha} \sum_{i=1}^{N} \alpha_i - \frac{1}{2} \sum_{i=1}^{N} \sum_{j=1}^{N} \alpha_i \alpha_j y_i y_j x_i x_j, \tag{12.30}$$

where N relates to the number of support vectors; L is the minimizing goal function in the formulation; α_i are the Lagrange multiplies x_i are the training attribute vectors; y_i relates to the training label vectors; and, b is the bias term of the hyperplane. The value of α_i is zero if it is not a support vector and nonzero if it is a support vector. The value of α_i can be found by solving Eq. (12.30) using quadratic programming (QP) techniques.

The optimal separating hyperplane is given by:

$$w = \sum_{i=1}^{N} \alpha_i y_i x_i \tag{12.31}$$

$$b = -\frac{1}{N_s} \sum_{s \in S} \left(y_s - \sum_{s \in S} \alpha_m x_m y_m \cdot x_s \right), \tag{12.32}$$

where S is a set of indices of the support vectors from each class. As discussed by Sanchez (2011), hard classification is obtained using the values of w, b according to the decision function in Eq. (12.26).

In any real Hilbert space, a hyperplane is composed of all the components, which satisfy the following expression:

$$\langle f, w \rangle_{\mathcal{H}} + c = 0 \tag{12.33}$$

The purpose of a traditional binary SVM classification task is to separate two different data classes $\mathcal{H}_+$ and $\mathcal{H}_-$ using a maximum margin hyperplane, as shown in Fig. 12.11.

In other words, elements belonging to the first class satisfy $\mathcal{H}_+ = \{f \in \mathcal{H}; \langle f, w \rangle_{\mathcal{H}} + c\} > 0$ true, whereas elements belonging to the second class satisfy $\mathcal{H}_- = \{f \in \mathcal{H}; \langle f, w \rangle_{\mathcal{H}} + c < 0\}$. As discussed earlier, in both of

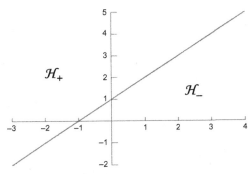

FIG. 12.11 Typical SVM hyperplane for linear classification.

these expressions, $w \in \mathcal{H}$, $b \in \mathbb{R}$ are the parameters that control the function and the decision rule is provided through Eq. (12.33).

Bouboulis and Theodoridis (2011) generalized the SVM formulations to complex spaces and subsequently extended them for other division 2 algebras such as quaternion classification (Bouboulis et al., 2015). In the case of complex valued data, the approach defines a complex space separated into four parts by using a pair of complex hyperplanes. In what follows, a notation where integers, real numbers, and complex numbers are denoted by $\mathbb{N}$, $\mathbb{R}$, and, $\mathbb{C}$, respectively, and $\bar{z}$ denotes the complex conjugate of z is adopted. A complex reproducing kernel Hilbert space (RKHS) will be denoted by $\mathbb{H}$, while a real RKHS by $\mathcal{H}$. The complex hyperplane is composed of two orthogonal hyperplanes that will be referred to as real and imaginary hyperplanes, as shown in Fig. 12.12.

The following expressions provides a complete description of the derived hyperplanes for some $w \in \mathbb{H}$, $c \in \mathbb{C}$, and $f \in \mathbb{H}$:

$$\mathrm{Re}\left(\langle f, w \rangle_{\mathbb{H}} + c\right) = 0 \tag{12.34a}$$

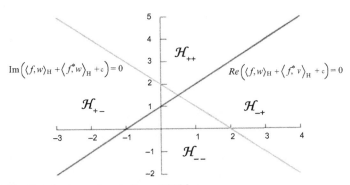

FIG. 12.12 Complex hyperplane in linear CSVM.

$$Im(\langle f, w \rangle_{\mathbb{H}} + c) = 0 \qquad (12.34b)$$

This is directly derived by observing that

$$\langle f, w \rangle_{\mathbb{H}} = \langle f^r, w^r \rangle_{\mathcal{H}} + \langle f^i, w^i \rangle_{\mathcal{H}} + i\left(\langle f^i, w^r \rangle_{\mathcal{H}} - \langle f^r, w^i \rangle_{\mathcal{H}}\right), \qquad (12.35)$$

where $f = f^r + if^i$ and $w = w^r + iw^i$.

The expressions naturally lead to the following conditions associated with the derivation of the hyperplanes:

$$\left\langle \begin{pmatrix} f^r \\ f^i \end{pmatrix}, \begin{pmatrix} w^r \\ w^i \end{pmatrix} \right\rangle_{\mathcal{H}^2} + b^r = 0 \qquad (12.36a)$$

$$\left\langle \begin{pmatrix} f^r \\ f^i \end{pmatrix}, \begin{pmatrix} -w^i \\ w^r \end{pmatrix} \right\rangle_{\mathcal{H}^2} + b^i = 0 \qquad (12.36b)$$

The expressions define two separate hyperplanes in $\mathcal{H}^2$. These planes are orthogonal if:

$$(-w^i w^r), \begin{pmatrix} -w^i \\ w^r \end{pmatrix} = 0 \qquad (12.37)$$

Moreover, in order to be able to define arbitrarily placed hyperplanes intersecting at oblique angles, widely linear estimation functions may also be employed. Through the expressions in Eq. (12.38), two hyperplanes associated with doubled real space such as $\mathcal{H}^2$ may be represented for some $v \in \mathbb{H}$, $c \in \mathbb{C}$ and $f \in \mathbb{H}$ arbitrarily on $\mathcal{H}^2$ depending on the values of w and v by:

$$Re(\langle f, w \rangle_{\mathbb{H}} + \langle f^*, v \rangle_{\mathbb{H}} + c) = 0 \qquad (12.38a)$$

$$Im(\langle f, w \rangle_{\mathbb{H}} + \langle f^*, v \rangle_{\mathbb{H}} + c) = 0, \qquad (12.38b)$$

where $w, v \in \mathbb{H}$, and $c \in \mathbb{C}$.

A corollary from explicitly adopting complex values leads to alternative expressions for the equations defining the hyperplanes in a complex space:

$$\left\langle \begin{pmatrix} f^r \\ f^i \end{pmatrix}, \begin{pmatrix} w^r + v^r \\ w^i - v^i \end{pmatrix} \right\rangle_{\mathcal{H}^2} + c^r = 0 \qquad (12.39a)$$

$$\left\langle \begin{pmatrix} f^r \\ f^i \end{pmatrix}, \begin{pmatrix} -(w^i + v^i) \\ w^r - v^r \end{pmatrix} \right\rangle_{\mathcal{H}^2} + c^i = 0 \qquad (12.39b)$$

Here,

$f = f^r + if^i$, $w = w^r + iw^i$, $v = v^r + iv^i$ and $c = c^r + ic^i$

In CSVM, a complex couple of hyperplanes can be defined as the set of $f \in \mathbb{H}$, which can satisfy any of these expressions (12.38a) or (12.38b), for some $w, v \in \mathbb{H}$ and $c \in \mathbb{C}$. As can be seen, the expressions (12.34)–(12.38) display the key difference between complex linear estimation and wide linear estimation functions. The complex linear case is pretty limiting, as the couple of complex hyperplanes is always orthogonal, while the widely linear case is broader and covers oblique hyperplane definitions. In CSVM, a complex pair of hyperplanes, therefore, separates the space of complex numbers $\mathbb{C}$ into four sectors as follows:

$$\mathcal{H}_{++} = \{f \in \mathcal{H}; \mathrm{Re}\left(\langle f, w\rangle_{\mathbb{H}} + \langle f^*, v\rangle_{\mathbb{H}} + c\right) > 0$$
$$\mathrm{Im}\left(\langle f, w\rangle_{\mathbb{H}} + \langle f^*, v\rangle_{\mathbb{H}} + c\right) > 0$$
$$\mathcal{H}_{+-} = \{f \in \mathcal{H}; \mathrm{Re}\left(\langle f, w\rangle_{\mathbb{H}} + \langle f^*, v\rangle_{\mathbb{H}} + c\right) > 0$$
$$\mathrm{Im}\left(\langle f, w\rangle_{\mathbb{H}} + \langle f^*, v\rangle_{\mathbb{H}} + c\right) < 0$$
$$\mathcal{H}_{-+} = \{f \in \mathcal{H}; \mathrm{Re}\left(\langle f, w\rangle_{\mathbb{H}} + \langle f^*, v\rangle_{\mathbb{H}} + c\right) < 0$$
$$\mathrm{Im}\left(\langle f, w\rangle_{\mathbb{H}} + \langle f^*, v\rangle_{\mathbb{H}} + c\right) > 0$$
$$\mathcal{H}_{--} = \{f \in \mathcal{H}; \mathrm{Re}\left(\langle f, w\rangle_{\mathbb{H}} + \langle f^*, v\rangle_{\mathbb{H}} + c\right) < 0$$
$$\mathrm{Im}\left(\langle f, w\rangle_{\mathbb{H}} + \langle f^*, v\rangle_{\mathbb{H}} + c\right) < 0 \tag{12.40}$$

In CSVM classification, the primary purpose is the identification of a complex couple of maximum margin hyperplanes that optimally (maximally) separate the points of the four classes, as can be seen in Fig. 12.12. For real and imaginary vectors w and v, the following expression is evaluated:

$$\left\|\begin{pmatrix} w^r + v^r \\ w^i - v^i \end{pmatrix}\right\|_{\mathcal{H}^2}^2 + \left\|\begin{pmatrix} -(w^i + v^i) \\ w^r - v^r \end{pmatrix}\right\|_{\mathcal{H}^2}^2$$
$$= \|w^r + v^r\|_{\mathcal{H}}^2 + \|w^i - v^i\|_{\mathcal{H}}^2 + \|w^i + v^i\|_{\mathcal{H}}^2 + \|w^r - v^r\|_{\mathcal{H}}^2 \tag{12.41}$$
$$= 2\|w^r\|_{\mathcal{H}}^2 + 2\|w^i\|_{\mathcal{H}}^2 + 2\|v^r\|_{\mathcal{H}}^2 + \|v^i\|_{\mathcal{H}}^2$$
$$= 2\left(\|w\|_{\mathbb{H}}^2 + \|v\|_{\mathbb{H}}^2\right)$$

Thus, the complex SVM optimization problem is defined as:

$$\min_{w,v,c} \frac{1}{2}\|w\|_{\mathbb{H}}^2 + \frac{1}{2}\|v\|_{\mathbb{H}}^2 + \frac{C}{N}\sum_{n=1}^{N}\left(\xi_n^r + \xi_n^i\right) \tag{12.42a}$$

subject to:

$$\begin{cases} d_n^r \mathrm{Re}\left(\Phi_{\mathbb{C}}(z_n)w_{\mathbb{H}} + \Phi_{\mathbb{C}}^*(z_n)v_{\mathbb{H}} + c\right) \geq 1 - \xi_n^r \\ d_n^i \mathrm{Im}\left(\Phi_{\mathbb{C}}(z_n)w_{\mathbb{H}} + \Phi_{\mathbb{C}}^*(z_n)w_{\mathbb{H}} + c\right) \geq 1 - \xi_n^i, \\ \xi_n^r, \xi_n^i \geq 0, \text{ for } n = 1, \ldots, N. \end{cases} \tag{12.42b}$$

where $\Phi_{\mathbb{C}}$ is the feature map, $Z = x + iy$

The associated Lagrangian function is then cast as follows:

$$
\begin{aligned}
L(w, v, a, b) =\ & \frac{1}{2}\|w\|_{\mathbb{H}}^2 + \frac{1}{2}\|v\|_{\mathbb{H}}^2 + \frac{C}{N}\sum_{n=1}^{N}\left(\xi_n^r + \xi_n^i\right) \\
& -\sum_{n=1}^{N} a_n\left(d_n^r \operatorname{Re}\left(\Phi_{\mathbb{C}}(z_n)w_{\mathbb{H}} + \Phi_{\mathbb{C}}^*(z_n)v_{\mathbb{H}} + c\right) - 1 - \xi_n^r\right) \\
& -\sum_{n=1}^{N} b_n\left(d_n^i \operatorname{Im}\left(\Phi_{\mathbb{C}}(z_n)w_{\mathbb{H}} + \Phi_{\mathbb{C}}^*(z_n)w_{\mathbb{H}} + c\right) - 1 + \xi_n^i\right) \\
& -\sum_{n=1}^{N}\eta_n\xi_n^r - \sum_{n=1}^{N}\theta_n\xi_n^i
\end{aligned}
\tag{12.43}
$$

Here, a_n, b_n, η_n, and θ_n are the positive Lagrange multipliers for $n=1,\ldots,N$, where N is the number of training examples and ξ_n^r, ξ_n^i are slack variables that permit margin failure. The constraints are given by the saddle point of the Lagrange functional equation and its minimum. The minimum of the Lagrange function in Eq. (12.43) occurs when:

$$
\frac{\partial L}{\partial w^*} = \frac{1}{2}w - \frac{1}{2}\sum_{n=1}^{N}a_n d_n^r \Phi_{\mathbb{C}}(z_n) + \frac{i}{2}\sum_{n=1}^{N}b_n d_n^i \Phi_{\mathbb{C}}(z_n)
\tag{12.44a}
$$

$$
\frac{\partial L}{\partial v^*} = \frac{1}{2}v - \frac{1}{2}\sum_{n=1}^{N}a_n d_n^r \Phi_{\mathbb{C}}^*(z_n) + \frac{i}{2}\sum_{n=1}^{N}b_n d_n^i \Phi_{\mathbb{C}}^*(z_n)
\tag{12.44b}
$$

$$
\frac{\partial L}{\partial c^*} = \frac{1}{2}\sum_{n=1}^{N}a_n d_n^r + \frac{i}{2}\sum_{n=1}^{N}b_n d_n^i
\tag{12.44c}
$$

$$
\frac{\partial L}{\partial \xi_n^r} = \frac{C}{N} - a_n - \eta_n
\tag{12.44d}
$$

$$
\frac{\partial L}{\partial \xi_n^i} = \frac{C}{N} - b_n - \theta_n
\tag{12.44e}
$$

when:

$$
\frac{\partial L}{\partial w} = 0 \Rightarrow w = \sum_{n=1}^{N}\left(a_n d_n^r - i b_n d_n^i\right)\Phi_{\mathbb{C}}(z_n)
\tag{12.45a}
$$

$$
\frac{\partial L}{\partial v} = 0 \Rightarrow v = \sum_{n=1}^{N}\left(a_n d_n^r - i b_n d_n^i\right)\Phi_{\mathbb{C}}^*(z_n)
\tag{12.45b}
$$

$$
\sum_{n=1}^{N}a_n d_n^r = \sum_{n=1}^{N}b_n d_n^i = 0
\tag{12.45c}
$$

$$\frac{\partial L}{\partial \xi_n^{er}} = 0 \Rightarrow a_n + \eta_n = \frac{C}{N} \tag{12.45d}$$

$$\frac{\partial L}{\partial \xi_n^{ei}} = 0 \Rightarrow b_n + \theta_n = \frac{C}{N} \tag{12.45e}$$

Using a Lagrangian, Eq. (12.40) can, therefore, be transformed into its dual form, which is a quadratic QP optimization problem. The following equation shows the result of using the conditions embodied in Eq. (12.45) with Eq. (12.43), (also known as the objective function of the dual problem). As can be seen, the dual problem in CSVM can be divided into two separate maximization tasks, as there are two hyperplanes, real and imaginary:

$$\underset{a}{\text{maximize}} \sum_{n=1}^{N} a_n - \sum_{n-m=1}^{N} a_n a_m d_n^r d_m^r \kappa_{\mathbb{C}}^r (z_m, z_n) \tag{12.46a}$$

subject to:

$$\begin{cases} \sum_{n=1}^{N} a_n d_n^r = 0 \\ 0 \leq a_n \leq \frac{C}{N} \end{cases} \tag{12.46b}$$

for $n = 1, \ldots, N$, and

$$\underset{a}{\text{maximize}} \sum_{n=1}^{N} b_n - \sum_{n,m=1}^{N} b_n b_m d_n^i d_m^i \kappa_{\mathbb{C}}^r (z_m, z_n) \tag{12.47a}$$

subject to:

$$\begin{cases} \sum_{n=1}^{N} b_n d_n^i = 0 \\ 0 \leq b_n \leq \frac{C}{N} \end{cases} \tag{12.47b}$$

for $n = 1, \ldots, N$. Here, N is equal to the number of support vectors, a_n is the Lagrange multiplier, and z the training attribute vector, where d_n^r is the real training label vector, d_n^i is the imaginary training label, and c the bias term of the hyperplane. Furthermore, a_n and b_n are Lagrange multipliers, with the value of both zero if it is not a support vector, and nonzero if it is a support vector. The value of both a_n and b_n can be found by solving Eqs. (12.46a) and (12.47a) using a QP-based SMO algorithm (Platt, 1998, 1999).

Once the Lagrange multipliers are determined, the real vector w, imaginary vector v, and the threshold c can be derived from the Lagrange multipliers. The hard classifier is obtained using the values of w, v, and c after placing them into the decision function as shown in Eq. (12.48a). In addition, the output of CSVM is explicitly computed from the Lagrange multipliers (a_n and b_n), as can be seen in Eq. (12.48b). Eq. (12.48) is a decision function, which is used to measure the performance of the classification and find the output of the classifier:

$$g(z) = \text{sign}_i \left(\langle \Phi_{\mathbb{C}}(z), w \rangle_{\mathbb{H}} + \langle \Phi_{\mathbb{C}}^*(z), v \rangle_{\mathbb{H}} + c \right) \tag{12.48a}$$

$$g(z) = \text{sign}_i \left(2 \sum_{n=1}^{N} \left(a_n d_n^r + i b_n d_n^i \right) \kappa_{\mathbb{C}}^r(z_n, z) + c^r + i c^i \right) \tag{12.48b}$$

$$\text{sign}_i(z) = \text{sign}[\text{Re}(z) + i \, \text{sign}(\text{Im}(z))] \tag{12.48c}$$

12.3.3   Recent advances in nonlinear multidimensional classifiers and state estimation algorithms

In many applications, using a linear SVM is not possible for obtaining accurate classification results due to the data not being linearly separable. In this case, a suitable mapping function needs to be used to transform the input data to a higher dimensional feature space where it can be linearly separable as required. A general nonlinear SVM can be expressed as:

$$f(x) = \sum_{i=1}^{N} \alpha_i y_i K(x_i, x_j) + b \tag{12.49}$$

The objective function of the dual problem for nonlinear SVM can be obtained on the basis of the following expression:

$$\max L(\alpha) = \max \sum_{i=1}^{N} \alpha_i - \frac{1}{2} \sum_{i=1}^{N} \sum_{j=1}^{N} \alpha_i \alpha_j y_i y_j K(x_i, x_j) \tag{12.50a}$$

subject to

$$\begin{cases} one \text{ linear equality constraint}: \ \sum_{i=1}^{N} \alpha_i y_i = 0 \\ \text{and the inequality constraints}: \ 0 \le \alpha_i \le C \end{cases} \tag{12.50b}$$

where $K(x_i, x_j)$ is a kernel functions that is used to measure the similarity of stored training samples to the input. Kernel functions can compute the dot product of two d-dimensional vectors in an implicitly higher dimensional space and transform the data from a low-dimensional input space to a higher-dimensional space. The most commonly used SVM kernels is the Gaussian radial basis function kernel:

$$K_\sigma, \mathbb{R}^v(x, y) = \exp \left(-\frac{\sum_{i=1}^{v} (x_i - y_i)^2}{\sigma^2} \right), \tag{12.51a}$$

$$K_{\sigma}, \mathbb{C}^{v}(x, y) = \exp\left(-\frac{\sum_{i=1}^{v}(z_i - w_i^*)^2}{\sigma^2}\right) \tag{12.51b}$$

where $x, y \in \mathbb{R}^{v}$, $z, w \in \mathbb{C}^{v}$, σ is a free positive parameter, related to the ith component of the complex vector $z \in \mathbb{C}^{v}$, and exp $(\bullet)$ is the extended exponential function in the complex domain (Bouboulis and Theodoridis, 2011). Other kernel functions such as polynomial or hyperbolic tangent can also be similarly extended to complex spaces. An application of the formulation can be found in the article by Jannah and Hadjiloucas (2015), which discusses ECG signal classification. As discussed in the article by Papageorgiou et al. (2015), the methodology is particularly useful for all types of nonlinear regression analysis.

Further to the algorithms for data classification, it must be added that processing in Reproducing Kernel Hilbert Spaces (RKHSs) in the context of online adaptive processing is also gaining popularity within the Signal Processing community (Kivinen et al., 2004; Engel et al., 2004; Slavakis et al., 2008, 2009; Slavakis and Theodoridis, 2008; Liu et al., 2008). The main advantage of RKHSs is that the original nonlinear task is "transformed" into a linear one. In addition, different types of nonlinearities can be treated in a unifying way, in a manner that does not affect the derivation of the algorithms, except at the final implementation stage. The procedure may be seen as being composed of two parts, firstly mapping the finite dimensionality input data from the input space into a higher dimensional space and then performing a linear processing (e.g., adaptive filtering) on the mapped data. When viewed collectively these two tasks are equivalent to a nonlinear processing (nonlinear filtering) procedure. Bouboulis and Theodoridis (2010) proposed a Complex Kernel Least-Mean-Square (CKLMS) algorithm that can be used to derive nonlinear stable algorithms, which offer significant performance improvements over the traditional complex LMS or Widely Linear complex LMS (WL-LMS) algorithms, when dealing with nonlinearities. This is particularly useful within the context of state estimation and of interest to the Kalman filter community.

A further development worth discussing here is the use of Clifford kernels instead of the division-2 algebraic kernels discussed. Clifford kernels may be used for classification or regression within an SVM context as well as for state estimation (Bayro-Corrochano and Daniel, 2010). In a Clifford algebra framework, the geometric product of two vectors $\mathbf{a}$ and $\mathbf{b}$ is defined as a sum of their inner product (symmetric part) and their wedge product (antisymmetric part)

$$\mathbf{ab} = \mathbf{a} \cdot \mathbf{b} + \mathbf{a} \wedge \mathbf{b} = \frac{1}{2}(\mathbf{ab} + \mathbf{ba}) + \frac{1}{2}(\mathbf{ab} - \mathbf{ba}) \tag{12.52}$$

The inner product of two vectors is the standard scalar or dot product and produces a scalar. The outer or wedge product of two vectors is called a bivector, an oriented area in the plane containing $\mathbf{a}$ and $\mathbf{b}$, formed by sweeping $\mathbf{a}$

along **b**. The outer product is immediately generalizable to higher dimensions, so a trivector $\mathbf{a} \wedge \mathbf{b} \wedge \mathbf{c}$ is interpreted as the oriented volume formed by sweeping the area $\mathbf{a} \wedge \mathbf{b}$ along vector $\mathbf{c}$. The outer product of k vectors is a k-vector or k-blade, and such a quantity is said to have grade k and a multivector $\mathbf{A}$ is the sum of k-blades of different or equal grade. For an n-dimensinal space $\mathbf{V}^n$, orthonormal basis vectors are introduced $\{\mathbf{e}_i\}$, $i = 1, \ldots n$ such that $e_i \cdot e_j = \delta_{ij}$, so that a basis that spans 1, $\{\mathbf{e}_i\}$, $\{\mathbf{e}_i \wedge \mathbf{e}_j\}$, $\{\mathbf{e}_i \wedge \mathbf{e}_j \wedge \mathbf{e}_k\}$, $\ldots \{\mathbf{e}_i \wedge \mathbf{e}_j \wedge \ldots \mathbf{e}_n\} = \mathbf{I}$ is generated, where $\mathbf{I}$ corresponds to the hypervolume associated with thte dataset (Hitzer, 2010; Bayro-Corrochano and Daniel, 2010). Applications of Clifford algebra in the Physical Sciences are discussed in Doran and Lasenby (2003), whereas its application to computer science are discussed in the work by Dorst et al. (2007). Clifford algebra-based biomedical image classification is discussed in the article by Yin et al. (2016). Since rotational and translational motion is naturally represented in this geometric framework, Euclidian motion is discussed in Dorst (2002), Dorst (2008). A MATLAB package (Geometric AlgeBra Learning Environment GABLE) has been developed in Waterloo (Mann et al., 1999) that may be used for such Clifford-based transformations. The application of Clifford algebra to state estimation and in particular within the context of Kalman filters is discussed by Bayro-Corrochano and Zhang (2000) who developed the motor extended Kalman filter (MEKF). This is a particularly important development that complements the algorithms discussed in the previous chapters of this book.

12.4 Recent advances in pattern recognition approaches for chemometrics

Classification of samples is one of the principal goals of pattern recognition. Methods for classification can be divided into unsupervised and supervised approaches. The difference between these methods is that for supervised approaches a test (or training) set is required: this means that certain samples of known origins and classification must first be analyzed to establish a model. In unsupervised methods, no prior test set is required.

The most commonly employed unsupervised method for classification is hierarchical cluster analysis, and there are a several computer packages available to perform this. The result is normally displayed graphically as a dendrogram. Objects that are most similar are joined together at the top levels of such a diagram. Clusters are often obvious to visualize. However, it is essential to recognize that there are a very large number of possible ways of computing dendrograms. Systematic approaches are necessary when applying any multivariate pattern recognition approach and considerations such as the method used for preprocessing are by no means unique to cluster analysis. However, this technique probably has the most complex set of options and so is a good example of the systematic choices encountered in chemometrics.

Binary tests for descriptions of color, texture, and so on often yield valuable information about the similarities between samples. Several coefficients have been proposed, including the Jaccard index, where the number of features common to two samples is divided by the total number of features present for both samples and the matching coefficient where the sum of features present and absent in both of two samples is divided by the sum of features present in only one of the two samples.

There are, however, a huge number of possible similarity measures. The most commonly used ones are the Euclidean distance or the Mahalanobis distance. Each sample can be represented as a point in an n-dimensional space (or n - 1 dimensions if the readings have been normalized), each dimension representing one variable. The straight line distance between each point is the geometric (Euclidean) distance. Another distance measure is the Manhattan or City block distance. This is the sum of the absolute distances along each axis. In two-dimensional space, the City block distance is the sum of the distances along the x and y axes of a point from the origin, whereas the Euclidean distance is the vector distance. A more generalized distance measure is the Minkowski distance. The City block distance is the one-dimensional Minkowski distance, the Euclidean distance the two-dimensional Minkowski distance; on this basis, three- and higher-dimensional distances can be computed. The Chebyshev distance is the longest side along the axis of the City block distance. An alternative approach is to use correlation coefficients. The nearer the correlation coefficient is to one, the closer samples are related. Correlation coefficients are measures of similarity, whereas distances are measures of dissimilarity. The Pearson r value is also often employed. An alternative is the cosine between the vectors for two points: at first sight this may not appear to be a correlation coefficient, but if two points are identical then the cosine between them is one. The latter similarity coefficient is a measure of direction rather than distance. Whether direction or distance is more relevant depends in part of how the original data are scaled.

Therefore, the choice of distance measure must depend on the questions being asked about the data and the associated preprocessing stages. A difficulty with the distance measures described here is that they take no account of possible correlations between the variables. For example, in pyrolysis several peaks might arise from a single compound. Compound A might yield ten peaks and compound B five peaks. Without any further knowledge of the system, compound A will contribute twice as much to the similarity matrix. This problem can be overcome by calculating the correlation coefficients between all the variables. The Mahalanobis distance measure takes this into account.

From the discussion, it follows that the choice of which distance space to use depends very much on what is known about the structure of the data. In particular, though, it is essential to remember how the data are scaled prior to choosing a distance measure. Further, it is important to consider whether variables are likely to be correlated (where Mahalanobis distance measures should be considered) or independent. If an inappropriate distance measure is used it is possible to obtain a meaningless output.

12.5   Recent advances in fuzzy methods for chemometrics

Another approach that has recently been applied to analytical data and is of relevance to chemometrics derives from fuzzy set theory. All the methods described in the previous sections are asumed to deal with matrices of exact numbers, where noise is modeled as a superimposed distribution on top of signals, and the actual numbers recorded by an instrument are assumed to be precise numbers. One has to accept, however, that all readings recorded by an instrument are subject to some error. A fuzzy model, however, that assigns a membership value between one and zero to each observation can be modeled by a membership function.

In most cases, these values can be modeled by Gaussians error distribution functions. If a reading is recorded as (x_1, y_1), where y is the response and x the dependent variable, then the point (x_1, y_1) is assigned a membership value of one. However, there is a finite probability that the true reading was, in fact, $(x_1 + \delta_{1x}, y_1 + \delta_{1y})$. This probability can be given by the membership function, which can be modeled as probability contours around the observed points. The criterion of the best fit straight line is normally given by minimizing a (weighted) least-square error estimator. In fuzzy theory, the criterion is obtained by maximizing the total membership function. The effectiveness of this approach has been demonstrated in linear calibration (Otto and Bandemer, 1986a).

The main weakness of least-squares methods is that they are strongly influenced by outliers that contribute disproportionally to the summed least-square estimator, whereas the fuzzy approach, which would set the membership function of an outlier close to zero, is normally more robust. Obviously, the effectiveness of this method depends somewhat on the prior model of the noise function and, as is usual for most chemometrics methods, improves if there is information in advance regarding the structure of the data. However, fuzzy methods take into account variability both in the x and y directions as noise is modeled by two- (or more) dimensional error space and so should be particularly relevant where there is uncertainty in sampling intervals. Fuzzy methods have also been compared with conventional multivariate approaches for classification of samples (Otto and Bandemer, 1986b). The efficiency of this approach depends, however, on the nature of the noise in the data and whether the fuzzy approaches correctly take into account the noise distribution.

From a state estimation perspective, when sensors are nonlinear, and there is also drift, the measurements are uncertain and the distribution of the noise is not symmetric. Therefore, it is not reasonable to describe the uncertainty in these systems as Gaussian distributions, which means that the EKF cannot estimate the states of these systems well. Some methods have been proposed on the basis of fuzzy logic and Kalman filter (Chao and Teng, 1996; Trajanoski and Wach, 1996; Chen et al., 1997, 1998; McGinnity and Irwin, 1997; Wu and Harris, 1997; Oussalah and Schutter, 2000; Longo et al., 2002; Simon, 2003; Matiá et al., 2006; Zhou et al., 2010). Among them, the extended fuzzy Kalman filter (EFKF) is an effective approach, accounting separately for the nonlinear system function and the nonlinear measurement function, respectively, by assigning fuzzy values (Zadeh, 1978) in the filter variables.

Finally, an interesting more recent development that uses the concept of Clifford algebra support vector machine classifiers in conjunction with fuzzy membership functions is discussed in the work by Wang et al. (2016). Their work addresses the problem that SVM is very sensitive to outliers (Guyon et al., 1996, Zhang (1999), Lin and Wang, 2002, Song, 2002), and in several applications training datasets may be corrupted by noise or accidentally misclassified.

From the foregoing discussions, it may be concluded that the use of appropriately chosen kernel functions that can linearize a problem, making use of Clifford algebras and fuzzy membership functions are important directions for advancing state estimation.

References

Adams, M., 1995. Chemometrics in Analytical Spectroscopy. The Royal Society of Chemistry, Cambridge, UK.

Alamo, T., Bravo, J.M., Camacho, E.F., 2005. Guaranteed state estimation by zonotopes. Automatica 41, 1035–1043.

Alsberg, B.K., Woodward, A.M., Kell, D.B., 1997. An introduction to wavelet transforms for chemometricians: a time-frequency approach. Chemom. Intell. Lab. Syst. 37, 215–239.

Alsberg, B.K., Woodward, A.M., Winson, M.K., Rowland, J.J., Kell, D.B., 1998. Variable selection in wavelet regression models. Anal. Chim. Acta 368, 29–44.

Araújo, M.C.U., Saldanha, T.C.B., Galvão, R.K.H., Yoneyama, T.H.C.C., Visani, V., 2001. The successive projections algorithm for variable selection in spectroscopic multicomponent analysis. Chemom. Intell. Lab. Syst. 57, 65–73.

Barnes, R., Shanoa, M., Lister, S., 1989. Standard normal variate transformation and de-trending of near-infrared diffuse reflectance spectra. Appl. Spectrosc. 43, 772–777.

Bayro-Corrochano, E.J., Daniel, N.A., 2010. Clifford support vector machines for classification, regression, and recurrence. IEEE Trans. Neural Netw. 21 (11), 1731–1746.

Bayro-Corrochano, E.J., Zhang, Y., 2000. The motor extended Kalman filter: a geometric approach for rigid motion estimation. J. Math. Imaging Vision 13, 205–228.

Beebe, K., Pell, R., Scasholtz, M.-B., 1998. Chemometrics, A Practical Guide. Wiley, New York.

Beezley, J.D., Mandel, J., Cobb, L., 2011. Wavelet ensemble Kalman filters. In: Proceedings of the 6th IEEE International Conference on Intelligent Data Acquisition and Advanced Computing Systems, Prague, Czech Republic, 15-17 Sept. 2011.

Bertsekas, D.P., Rhodes, I.B., 1971. Recursive state estimation for a set-membership description of uncertainty. IEEE Trans. Autom. Control 16 (2), 117–128.

Borgen, O., Kowalski, B., 1985. An extension of the multivariate component-resolution method to three components. Anal. Chim. Acta 174, 1–26.

Bouboulis, P., Theodoridis, S., 2010. Extension of Wirtinger calculus in RKH spaces and the Complex Kernel LMS. In: 2010 IEEE International Workshop on Machine Learning for Signal Processing, Kittilapp. 136–141.

Bouboulis, P., Theodoridis, S., 2011. Extension of Wirtinger 's Calculus in reproducing kernel Hilbert spaces and the complex kernel LMS. IEEE Trans. Signal Process. 59 (3), 964–978.

Bouboulis, P., Theodoridis, S., Mavroforakis, C., Evaggelatou-Dalla, L., 2015. Complex support vector machines for regression and quaternary classification. IEEE Trans. Neural Netw. Learn. Syst. 26 (6), 1260–1274.

Brereton, R.G., 1987. Chemometrics in analytical chemistry a review. Analyst 112, 1635–1657.

Brereton, R., 2003. Chemometrics, Data Analysis for the Laboratory and Chemical Plant. Wiley, Chichester.

Burges, C.J.C., 1998. A tutorial on support vector machine for pattern recognitionx. Data Min. Knowl. Disc. 2, 121–167.

Carlin, B.P., Polson, N.G., Stoffer, D.S., 1992. A Monte Carlo approach to nonnormal and nonlinear state-space modeling. J. Am. Stat. Assoc. 87, 493–500.

Carroll, J.B., 1957. Biquartimin criterion for rotation to oblique simple structure in factor analysis. Science 126, 1114–1115.

Caujolle, M., Petit, M., Fleury, G., Berthet, L., 2010. Reliable power disturbance detection using wavelet decomposition or harmonic model based Kalman filtering. In: Proceedings of 14th International Conference on Harmonics and Quality of Power—ICHQP 2010, Bergamopp. 1–6.

Chabane, S.B., Stoica, M.C., Alamo, T., Camacho, E.F., Dumur, D., 2014. A new approach for guaranteed ellipsoidal state estimation. In: Proceedings of the 19th World Congress the International Federation of Automatic Control, Cape Town, South Africa, August 24–29pp. 6533–6538.

Chao, C.T., Teng, C.C., 1996. A fuzzy neural network based extended Kalman filter. Int. J. Syst. Sci. 27, 333–339.

Chen, R., Liu, J.S., 2000. Mixture Kalman filters. J. R. Stat. Soc. Ser. B 62, 493–508.

Chen, J., Wang, X.Z., 2001. A new approach to near-infrared spectral data analysis using independent component analysis. J. Chem. Inf. Comput. Sci. 41, 992–1001.

Chen, G., Wang, J., Shieh, L.S., 1997. Interval Kalman filtering. IEEE Trans. Aerosp. Electron. Syst. 33, 251–259.

Chen, G., Xie, Q., Shieh, L.S., 1998. Fuzzy Kalman filtering. Inf. Sci. 109, 197–209.

Chen, Q., Krunger, U., Meronk, M., Leung, A., 2004. Synthesis of t^2 and q statistics for process monitoring. Control. Eng. Pract. 12, 745–755.

Chernousko, F.L., 1994. State Estimation for Dynamic Systems. CRC Press, Boca Raton.

Chisci, L., Garulli, A., Zappa, G., 1996. Recursive state bounding by parallelotopes. Automatica 32, 1049–1055.

Chou, C.-M., Wang, R.-Y., 2004. Application of wavelet-based multi-model Kalman filters to real-time flood forecasting. Hydrol. Process. 18, 987–1008.

Cichocki, A., Amari, S., 2002. Adaptive Blind Signal and Image Processing: Learning Algorithms and Applications. Wiley.

Collin, F., Warnant, R., 1995. Application of the wavelet transform for GPS cycle slip correction and comparison with Kalman Filter. Manuscr. Geodaet. 20, 161–172.

Combastel, C., 2003. A state bounding observer based on zonotopes. In: European Control Conference (ECC), Cambridge, UKpp. 2589–2594.

Croux, C., Ruiz-Gazen, A., 1996. A fast algorithm for robust principal components based on projection pursuit. In: COMPSTAT: Proceedings in Computational Statistics. Physica-Verlag, Heidelberg, pp. 211–217.

Cuesta-Sanchez, F., Toft, J., van den Bogaert, B., Massart, D., 1996. Orthogonal projection approach (OPA) applied to peak purity assessment. Anal. Chem. 68, 79–85.

Daryin, A.N., Kurzhanski, A.B., 2012. Estimation of reachability sets for large-scale uncertain systems: from theory to computation. In: Proceedings of the of 51st IEEE Conference on Decision and Control, Maui, Hawaii, USA.pp. 7401–7406.

Daryin, A.N., Kurzhanski, A.B., Vostrikov, I.V., 2006. Reachability approaches and ellipsoidal techniques for closed-loop control of oscillating systems under uncertainty. In: Proceedings of the of 51st IEEE Conference on Decision and Control, San Diego, CA, USA.pp. 6390–6395.

Daszykowski, M., 2007. From projection pursuit to other unsupervised chemometric techniques. J. Chem. Aust. 21, 270–279.

Daubechies, I., 1992. Ten Lectures on Wavelets. CBMS-NSF Regional Conference Series in Applied Mathematics, vol. 61. SIAM, Philadelphia, PA.

de Jong, S., Kiers, H., 1992. Principal covariates regression. Part 1, theory. Chemom. Intell. Lab. Syst. 14, 155–164.

De Lathauwer, L., De Moor, B., Vandewalle, J., 2000. An introduction to independent component analysis. J. Chemom. 14, 123–149.

Doran, C., Lasenby, A., 2003. Geometric Algebra for Physicists. Cambridge University Press, Cambridge, UK.

Dorst, L., 2002. Tutorial: structure-preserving representation of euclidean motions through conformal geometric algebra computing. IEEE Comput. Graph. Appl. 22 (3), 24–31.

Dorst, L., 2008. The representation of rigid body motions in the conformal model of geometric algebra. In: Rosenhahn, B., Klette, R., Metaxas, D. (Eds.), Human Motion. Springer, Dordrecht, The Netherlands, pp. 507–529.

Dorst, L., Fontijne, D., Mann, S., 2007. Geometric Algebra for Computer Science: An Object-Oriented Approach to Geometry. Morgan Kaufmann, Burlington, MA.

Draper, N., Smith, H., 1981. Applied Regression Analysis, second ed. Wiley, New York.

Durieu, C., Walter, E., Polyak, B., 2001. Multi-input multi-output ellipsoidal state bounding. J. Optim. Theory Appl. 111 (2), 273–303.

Engel, Y., Mannor, S., Meir, R., 2004. The kernel recursive least-squares algorithm. IEEE Trans. Signal Process. 52 (8), 2275–2285.

Fei, B., Liu, J., 2006. Binary tree of SVM: a new fast multiclass training and classification algorithm. IEEE Trans. Neural Netw. 17 (3), 696–704.

Fogel, E., Huang, Y.F., 1982. On the value of information in system identification-bounded noise case. Automatica 18, 229–238.

Friedman, J.H., Tuckey, J.W., 1974. A projection pursuit algorithm for exploratory data analysis. IEEE Trans. Comput. C-23 (9), 881–890.

Froese, T., Hadjiloucas, S., Galvao, R.K.H., Becerra, V.M., Coelho, C.J., 2006. Comparison of extrasystolic ECG signal classifiers using discrete wavelet transforms. Pattern Recogn. Lett. 27 (5), 393–407.

Galvão, R.K.H., Hadjiloucas, S., Bowen, J.W., 2002. Use of the statistical properties of the wavelet transform coefficients for the optimisation of integration time in Fourier transform spectrometry. Opt. Lett. 27, 643–645.

Galvão, R.K.H., Hadjiloucas, S., Bowen, J.W., Coelho, C.J., 2003. Optimal discrimination and classification of THz spectra in the wavelet domain. Opt. Express 11 (12), 1462–1473.

Galvão, R.K.H., Jose, G.E., Dantas Filho, H.A., Araujo, M.C.U., Silva, E.C., Paiva, H.M., Saldanha, T.C.B., 2004. Wavelet optimization exploiting X and Y data statistics. Chemom. Intell. Lab. Syst. 70, 1–10.

Galvão, R.K.H., Hadjiloucas, S., Zafiropoulos, A., Walker, G.C., Bowen, J.W., Dudley, R., 2007a. Optimization of apodization functions in terahertz transient spectrometry. Opt. Lett. 32 (20), 3008–3010.

Galvão, R.K.H., Filho, H.A.D., Martins, M.N., Araújo, M.C.U., Pasquini, C., 2007b. Sub-optimal wavelet denoising of coaveraged spectra employing statistics from individual scans. Anal. Chim. Acta 581, 159–167.

Galvão, R.K.H., Hadjiloucas, S., Kienitz, K.H., Paiva, H.M., Afonso, R.J.M., 2013a. Fractional order modeling of large three-dimensional RC networks. IEEE Trans. Circuits Syst. I 60 (3), 624–637.

Galvão, R.K.H., Kienitz, K.H., Hadjiloucas, S., Walker, G.C., Bowen, J.W., Soares, S.F.C., Araújo, M.C.U., 2013b. Multivariate analysis of random three-dimensional RC networks in the time and frequency domains. IEEE Trans. Dielectr. Electr. Insul. 20 (3), 995–1008.

Galvão, R.K.H., Hadjiloucas, S., Matsuura, J.P., Colombo Jr., J.R., 2017. A model-based approach for screening analysis of compositional changes in three-dimensional RC networks. IEEE Trans. Dielectr. Electr. Insul. 24 (2), 1141–1152.

Galvão, R.K.H., Kienitz, K.H., Hadjiloucas, S., 2018. Conversion of descriptor representations of electrical circuits to state-space form: an extension of the shuffle algorithm. Int. J. Control. 91 (10), 2199–2213.

Garzelli, A., Nencini, F., 2007. Panchromatic sharpening of remote sensing images using a multi-scale Kalman filter. Pattern Recogn. 40 (12), 3568–3577.

Geladi, P., 2001. Some recent trends in the calibration literature. Chemom. Intell. Lab. Syst. 60, 211–224.

Geladi, P., Dábakk, E., 1995. An overview of chemometrics applications in NIR spectrometry. J. Near Infrared Spectrosc. 3, 119–132.

Geladi, P., Mc, D.D., Martens, H., 1985. Linearization and scatter-correction for near-infrared reflectance spectra of meat. Appl. Spectrosc. 39, 491–500.

Gemperline, P., 1986. Target transformation factor analysis with linear inequality constraints applied to spectroscopic–chromatographic data. Anal. Chem. 58, 2656–2663.

Geweke, J., Tanizaki, H., 1999. On Markov chain Monte Carlo methods for nonlinear and non-Gaussian state-space models. Commun. Stat. Simul. Comput. 28, 867–894.

Glover, D., Hopke, P.K., 1992. Exploration of multivariate chemical data by projection pursuit. Chemom. Intell. Lab. Syst. 16, 45–59.

Golightly, A., Wilkinson, D.J., 2005. Bayesian inference for stochastic kinetic models using a diffusion approximation. Biometrics 61, 781–788.

Guyon, I., Matic, N., Vapnik, V.N., 1996. Discovering Information Patterns and Data Cleaning. MIT Press, Cambridge, pp. 181–203.

Hadjiloucas, S., Galvão, R.K.H., Bowen, J.W., 2002. Analysis of spectroscopic measurements of leaf water content at THz frequencies using linear transforms. J. Opt. Soc. Am. A 19, 2495–2509.

Hadjiloucas, S., Galvão, R.K.H., Becerra, V.M., Bowen, J.W., Martini, R., Brucherseifer, M., Pellemans, H.P.M., Bolivar, P.H., Kurz, H., Chamberlain, J.M., 2004. Comparison of subspace and ARX models of a waveguide's terahertz transient response after optimal wavelet filtering. IEEE Trans. Microwave Theory Tech. 52 (10), 2409–2419.

Hadjiloucas, S., Walker, G.C., Bowen, J.W., Becerra, V.M., Zafiropoulos, A., Galvão, R.K.H., 2009. High signal to noise ratio THz spectroscopy with ASOPS and signal processing schemes for mapping and controlling molecular and bulk relaxation processes. J. Phys. Conf. Ser. 183, 012003.

Hadjiloucas, S., Jannah, N., Hwang, F., Galvão, R.K.H., 2014. On the application of optimal wavelet filter banks for ECG signal classification. J. Phys. Conf. Ser. 490 (1), 012142.

Harman, H.H., 1967. Modern Factor Analysis, third ed. University of Chicago Press, Chicago.

Hearmon, R.A., Scrivens, J.H., Jennings, K.R., Farncombe, M.J., 1987. Isolation of component spectra in the analysis of mixtures by mass spectrometry and ^{13}C nuclear magnetic resonance spectroscopy: the utility of abstract factor analysis. Chemom. Intell. Lab. Syst. 1, 167–176.

Helland, I., 1988. On the structure of partial least squares regression. Commun. Stat. B Simul. Comput. 17, 581–607.

Hitzer, E., 2010. Angles between subspaces computed in clifford algebra. In: Proceeding of International Conference on Numerical Analysis and Applied Mathematics.vol. 1281. pp. 1476–1479.

Ho, C., Christian, G., Davidson, E., 1978. Application of the method of rank annihilation to quantitative analyses of multicomponent fluorescence data from the video fluorometer. Anal. Chem. 50, 1108–1113.

Hoerl, A., Kennard, R., 1970. Ridge regression: biased estimation for nonorthogonal problems. Technometrics 8, 27–51.

Horst, P., 1965. Factor Analysis of Data Matrices. Holt, Rinehart and Winston, New York.

Hoskuldsson, A., 1988. PLS regression methods. J. Chem. Aust. 2, 211–228.

Hu, T., Lin, Z., 2003. Composite quadratic Lyapunov functions for constrained control systems. IEEE Trans. Autom. Control 48, 440–450.

Hyvärinen, A., Oja, E., 2000. Independent component analysis: algorithms and applications. Neural Netw. 13, 411–430.

Hyvärinen, A., Karhunen, J., Oja, E., 2001. Independent Component Analysis. John Willey & Sons, Inc., New York

Jacyntho, L.A., Teixeira, M.C.M., Assunção, E., Cardim, R., Galvão, R.K.H., Hadjiloucas, S., 2015. Identification of fractional-order transfer functions using a step excitation. IEEE Trans. Circuits Syst. II Express Briefs 62, 896–900.

Jannah, N., Hadjiloucas, S., 2015. Detection of ECG arrhythmia conditions using CSVM and MSVM classifiers. In: 2015 IEEE Signal Processing in Medicine and Biology Symposium (SPMB), Philadelphia, PApp. 1–2.

Joachims, T., 1998. Making large scale SVM learning practical. In: Scholkopf, B., Bruges, C., Smola, A. (Eds.), Advances in Kernel methods-support vector learning. MIT Press, Cambridge, MA.

Johnson, A.E.W., Cholleti, S.R., Buchman, T.G., Clifford, G.D., 2013. Improved respiration rate estimation using a Kalman filter and wavelet cross-coherence. In: Computing in Cardiology. pp. 791–794Zaragoza.

Kaiser, H.F., 1958. The varimax criterion for analytic rotation in factor analysis. Psychometrika 23, 187–200.

Karjalainen, E.J., Karjalainen, U.P., 1996. Data Analysis for Hyphenated Techniques. Elsevier, Amsterdam, Netherlands.

Kemsley, E., 1998. Discriminant Analysis and Class Modelling of Spectroscopic Data. Wiley, Chichester.

Kim, K.J., Park, J.B., Choi, Y.H., 2006. The adaptive learning rates of extended Kalman Filter based training algorithm for wavelet neural networks. In: MICAI 2006: Advances in Artificial Intelligence.pp. 327–337.

Kivinen, J., Smola, A., Williamson, R.C., 2004. Online learning with kernels. IEEE Trans. Signal Process. 52 (8), 2165–2176.

Kurzhanski, A.B., Vályi, I., 1996. Ellipsoidal Calculus for Estimation and Control. Birkhaüser, Boston.

Le, V.T.H., Stoica, C., Alamo, T., Camacho, E.F., Dumur, D., 2013. Zonotopic guaranteed state estimation for uncertain systems. Automatica 49 (1), 3418–3424.

Li, K.-C., 1991. Sliced inverse regression for dimension reduction. J. Am. Stat. Assoc. 86, 316–342.

Liang, Y., Kvalheim, O., Rahmani, A., Brercton, R., 1993. Resolution of strongly overlapping two-way multicomponent data by means of heuristic evolving latent projections. J. Chemom. 7, 15–43.

Lin, C., Wang, S., 2002. Fuzzy support vector machines. IEEE Trans. Neural Netw. 13 (2), 464–471.

Liu, W., Pokharel, P., Principe, J.C., 2008. The kernel least mean-square algorithm. IEEE Trans. Signal Process. 56 (2), 543–554.

Longo, D., Muscato, G., Sacco, V., 2002. Localization using Fuzzy and Kalman filtering data fusion. In: Fifth International Conference on Climbing and Walking Robots. CLAWAR, Paris, France, pp. 263–270.

Maeder, M., 1987. Evolving factor analysis for the resolution of overlapping chromatographic peaks. Anal. Chem. 59, 527–530.

Malinowski, E.R., 1987. Theory of the distribution of error eigenvalues resulting from principal component analysis with applications to spectroscopic data. J. Chem. Aust. 1, 33–40.

Malinowski, E.R., Howery, D.G., 1980. Factor Analysis in Chemistry. Wiley, New York.

Mann, S., Dorst, L., Bouma, T., 1999. The making of a geometric algebra package in matlab. In: Research Report CS-99-27. University of Waterloo.

Martens, H., Martens, M., 2000. Multivariate Analysis of Quality, An Introduction. Wiley, Chichester.

Martens, H., Næcs, T., 1989. Multivariate Calibration. Wiley, Chichester.

Massy, W., 1965. Principal components regression in exploratory statistical research. J. Am. Stat. Assoc. 60, 234–256.

Matiá, F., Jiménez, A., Al-Hadithi, B.M., Rodríguez-Losada, D., Galán, R., 2006. The fuzzy Kalman filter: state estimation using Possibilistic techniques. Fuzzy Sets Syst. 157, 2145–2170.

McGinnity, S., Irwin, G., 1997. Nonlinear state estimation using fuzzy local linear models. Int. J. Syst. Sci. 28, 643–656.

Næcs, T., Isaksson, T., 1989. Selection of samples for calibration in near-infrared spectroscopy. Part 1. General principles illustrated by example. Appl. Spectrosc. 43, 328–335.

Niemi, J., West, M., 2010. Adaptive mixture modeling Metropolis methods for Bayesian analysis of nonlinear state-space models. J. Comput. Graph. Stat. 19 (2), 260–280.

Odiowei, P.P., Cao, Y., 2010. State-space independent component analysis for nonlinear dynamic process monitoring. Chemom. Intell. Lab. Syst. 103, 59–65.

Otto, M., Bandemer, H., 1986a. Pattern recognition based on fuzzy observations for spectroscopic quality control and chromatographic fingerprinting. Anal. Chim. Acta 184, 21–31.

Otto, M., Bandemer, H., 1986b. Calibration with imprecise signals and concentrations based on fuzzy theory. Chemom. Intell. Lab. Syst. 1, 71–78.

Oussalah, M., Schutter, J.D., 2000. Possibilistic Kalman filtering for radar 2D tracking. Inf. Sci. 130, 85–107.

Paiva, H.M., Galvão, R.K.H., 2006. Wavelet-packet identification of dynamic systems in frequency subbands. Signal Process. 86, 2001–2008.

Paiva, J.P.L.M., Kelencz, C.A., Paiva, H.M., Galvão, R.K.H., Magini, M., 2008. Adaptive wavelet EMG compression based on local optimization of filter banks. Physiol. Meas. 29, 843–856.

Paiva, H.M., Martins, M.N., Galvão, R.K.H., Paiva, J., 2009. On the space of orthonormal wavelets: additional constraints to ensure two vanishing moments. IEEE Signal Process Lett. 6, 101–104.

Papageorgiou, G., Bouboulis, P., Theodoridis, S., 2015. Robust non-linear regression analysis: a greedy approach employing kernels. J. Mach. Learn. Res. 1, 1–48.

Platt, J.C., 1998. Sequential minimal optimization: a fast algorithm for training support vector machines. In: Advances in Kernel Methods Support Vector Learning.

Platt, J., 1999. Fast training of support vector machines using sequential minimal optimization. In: Advances in Kernel Methods—Support Vector Learning. MIT Press, Cambridge, MA, pp. 185–208.

Platt, J., Cristianini, N., Shawe-Taylor, J., 2000. Large margin DAGSVM's for multiclass classification. Adv. Neural Inf. Proces. Syst. 12, 547–553.

Polyak, B., Nazin, S.A., Durieu, C., Walter, E., 2004. Ellipsoidal parameter or state estimation under model uncertainty. Automatica 40, 1171–1179.

Postalcloglu, S., Erkan, K., Bolat, E.D., 2005. Comparison of Kalman filter and Wavelet filter for denoising. In: 2005 International Conference on Neural Networks and Brain, Beiging Chinapp. 951–954.

Puig, V., Cugueró, P., Quevedo, J., 2001. Worst-case state estimation and simulation of uncertain discrete-time systems using zonotopes. In: 2001 European Control Conference (ECC), Portopp. 1691–1697.

Renaud, O., Starck, J.-L., Murtagh, F., 2005. Wavelet-based combined signal filtering and prediction. IEEE Trans. Syst. Man Cybern. B 35 (6), 1241–1251.

Rosenfeld, N., Young, J., Alon, U., Swain, P., Elowitz, M., 2005. Gene regulation at the single-cell lesvel. Science 307, 1962–1965.

Rummel, R.J., 1970. Applied Factor Analysis. Northwestern University Press, Evanston, pp. 372–385.

Salam, F.M., Waheed, K., 2001. State space feedforward and feedback structures for blind source recovery. In: 3rd International Conference on ICA and BSS, December 9–12, San Diego, CApp. 248–253.

Salis, C.I., Malissovas, A.E., Bizopoulos, P.A., Tzallas, A.T., Angelidis, P.A., Tsalikakis, D.G., 2013. Denoising simulated EEG signals: a comparative study of EMD, wavelet transform and Kalman filter. In: 13th IEEE International Conference on BioInformatics and BioEngineering, 10–13 Nov. 2013, Chania, Greece.

Sanchez, G.G., 2011. Examination of the applicability of support vector machines in the context of ischaemia detection. In: UPCs New Academic and Scientific Communications. Dresden University, pp. 1–95.

Sanchez, E., Kowalski, B.R., 1986. Generalized rank annihilation factor analysis. Anal. Chem. 58, 496–499.

Saunders, D.R., 1960. A computer program to find the best-fitting orthogonal factors for a given hypothesis. Psychometrika 25, 199–205.

Schweppe, F.C., 1968. Recursive state estimation: unknown but bounded errors and system inputs. IEEE Trans. Autom. Control 13 (1), 22–28.

Sherlock, B.G., Monro, D.M., 1998. On the space of orthonormal wavelets. IEEE Trans. Signal Process. 46, 1716–1720.

Simon, D., 2003. Kalman filtering for fuzzy discrete time dynamic systems. Appl. Soft Comput. 3, 191–207.

Slavakis, K., Theodoridis, S., 2008. Sliding window generalized kernel affine projection algorithm using projection mappings. Eur. J. Adv. Signal Process. 2008, 735351, 16 pages.

Slavakis, K., Theodoridis, S., Yamada, I., 2008. On line classification using kernels and projection based adaptive algorithm. IEEE Trans. Signal Process. 56 (7), 2781–2797.

Slavakis, K., Theodoridis, S., Yamada, I., 2009. Adaptive constrained learning in reproducing kernel hilbert spaces: the robust beamforming case. IEEE Trans. Signal Process. 57 (12), 4744–4764.

Song, Q., 2002. Robust support vector machine with bullet hole image classification. IEEE Trans. Syst. Cybern. 32 (4), 440–448.

Stone, M., Brooks, R., 1990. Continuum regression: crossvalidated sequentially constructed prediction embracing ordinary least squares, partial least squares and principal component regression. J. R. Stat. Soc. B 52, 237–269.

Strang, G., Nguyen, T., 1998. Wavelets and Filter Banks. Wellesley-Cambridge, Wellesley, MA.

Stroud, J.R., Müller, P., Polson, N.G., 2003. Nonlinear state-space models with state-dependent variances. J. Am. Stat. Assoc. 98, 377–386.

Sudarsanam, A., Barnes, R., Carver, J., Kallam, R., Dasu, A., 2010. Dynamically reconfigurable systolic array accelerators: a case study with extended Kalman filter and discrete wavelet transform algorithms. IET Comput. Digit. Tech. 4 (2), 126–142.

Svensson, O., Kourti, T., MacGregor, J., 2002. An investigation of orthogonal signal correction algorithms and their characteristics. J. Chem. Aust. 16, 176–188.

Tan, C., Song, H., Niemi, J., You, L., 2007. A synthetic biology challenge: making cells compute. Mol. BioSyst. 3, 343–353.

Tangborn, A., Zhang, S.Q., 2000. Wavelet transform adapted to an approximate Kalman filter system. Appl. Numer. Math. 33 (1–4), 307–316.

Tauler, R., Barcelo, D., 1993. Multivariate curve resolution applied to liquid chromatography diode array detection. Trends Anal. Chem. 12, 319–327.

Trajanoski, Z., Wach, P., 1996. Fuzzy filter for state estimation of a Glucoregulatory system. Comput. Methods Prog. Biomed. 50, 265–273.

Tuqun, J., Vaidyanathan, P.P., 2000. A state-space approach to the design of globally optimal FIR energy compaction filters. IEEE Trans. Signal Process. 48, 2822–2838.

Uzinski, J.C., Paiva, H.M., Alvarado, F.V., Duarte, M.A.Q., Galvão, R.K.H., 2013. Additional Constraints to Ensure Three Vanishing Moments for Orthonormal Wavelet Filter Banks Anais do Congresso de Matemática Aplicada e Computacional CMAC Centro-Oeste. vol. 2013. pp. 16–19.

Uzinski, J.C., Paiva, H.M., Galvão, R.K.H., Assunção, E., Duarte, M.A.Q., Villarreal, F., 2017. A dynamic-state feedback approach employing a new state-space description for the FastWavelet transform with multiple decomposition levels. J. Control Autom. Electr. Syst. 28 (3), 303–313.

Vaidyanathan, P.P., 1993. Multirate Systems and Filter Banks. Prentice-Hall, Englewood Cliffs, NJ.

Vandeginste, W., Derks, G.K., 1985. Multicomponent self-modeling curve resolution in high-performance liquid chromatography by iterative target transformation analysis. Anal. Chim. Acta 173, 253–264.

Vapnik, V., 1998. Statistical Learning Theory. Wiley, New York.

Vapnik, V., 1999. The Nature of Statistical Learning Theory, second ed. Springer, New York.

Visser, E., Lee, T., 2004. An information-theoretic methodology for the resolution of pure component spectra without prior information using spectroscopic measurements. Chemom. Intell. Lab. Syst. 70, 147–155.

Waheed, K., Salam, F.M., 2003. Blind source recovery: a framework in the state space. J. Mach. Learn. Res. 4, 1411–1446.

Walter, E., Piet-Lahanier, H., 1989. Exact recursive polyhedral description of the feasible parameter set for bounded-error models. IEEE Trans. Autom. Control 34 (8), 911–915.

Wang, G., Cai, W., Shao, X., 2006. A primary study on resolution of overlapping GC-MS signal using mean-field approach independent component analysis. Chemom. Intell. Lab. Syst. 82, 137–144.

Wang, Q., Niemi, J., Tan, C., You, L., West, M., 2010. Image segmentation and dynamic lineage analysis in single-cell fluorescent microscopy. Cytometry A 77, 101–110.

Wang, R., Zhang, X., Cao, W., 2016. Clifford fuzzy support vector machines for classification. Adv. Appl. Clifford Algebr. 26, 825–846.

Webster, J., Gunst, R., Mason, R., 1974. Latent root regression analysis. Technometrics 16, 513–522.

West, M., Harrison, J., 1997. Bayesian Forecasting and Dynamic Models, second ed. SpringerVerlag, New York.

Westad, F., 2005. Independent component analysis and regression applied on sensory data. J. Chem. Aust. 19, 171–179.

Westad, F., Kermit, M., 2003. Cross validation and uncertainty estimates in independent component analysis. Anal. Chim. Acta 490, 341–354.

Weston, J., Watkins, C., 1999. Multi-class support vector machines. In: Verleysen, M. (Ed.), Proceedings of ESANN99.Brussels, Belgium.

Wilkinson, D., 2006. Stochastic Modelling for Systems Biology. Chapman & Hall/CRC, London.

Windig, W., Guilment, J., 1991. Interactive self-modeling mixture analysis. Anal. Chem. 63, 1425–1432.

Wistenhausen, H.S., 1968. Sets of possible states of linear systems given perturbed observations. IEEE Trans. Autom. Control 13 (5), 556–558.

Wold, S., 1978. Cross-validatory estimation of the number of components in factor and principal components models. Technometrics 20, 397–405.

Wu, Z.Q., Harris, C.J., 1997. A Neurofuzzy network structure for modelling and state Estimacion of unknown nonlinear systems. Int. J. Syst. Sci. 28, 335–345.

Xie, Y., Zhang, Y., Ye, Z., 2007. Short-term traffic volume forecasting using Kalman filter with discrete wavelet decomposition. Comput. Aided Civ. Inf. Eng. 22, 326–334.

Yin, X., Ng, B.W.-H., Ferguson, B., Abbott, D., Hadjiloucas, S., 2007. Auto-regressive models of wavelet sub-bands for classifying terahertz pulse measurements. J. Biol. Syst. 15 (4), 551–571.

Yin, X.X., Hadjiloucas, S., Fischer, B.M., Ng, B.W.H., Paiva, H.M., Galvao, R.K.H., Walker, G.C., Bowen, J.W., Abbott, D., 2008. Classification of lactose and mandelic acid THz spectra using subspace and wavelet-packet algorithms. In: Proceedings of the SPIE 6798, Microelectronics: Design, Technology, and Packaging IIIp. 679814.

Yin, X.-X., Zhang, Y., Cao, J., Wu, J.-L., Hadjiloucas, S., 2016. Exploring the complementarity of THz pulse imaging and DCE-MRIs: toward a unified multi-channel classification and a deep learning framework. Comput. Methods Prog. Biomed. 137, 87–114.

Yin, X.-X., Hadjiloucas, S., Zhang, Y., 2017. Pattern Classification of Medical Images: Computer Aided Diagnosis. Springer International Publishing.

Zadeh, L.A., 1978. Fuzzy sets as a basis for a theory of possibility. Fuzzy Sets Syst. 1, 3–28.

Zhang, X., 1999. Using class-center vectors to build support vector machines. In: Neural Networks for Signal Processing IX: Proceedings of the 1999 IEEE Signal Processing Society Workshop (Cat. No.98TH8468), Madison, WI, USApp. 3–11.

Zheng, T., Girgis, A.A., Makram, E.B., 2000. A hybrid wavelet-Kalman filter method for load forecasting. Electr. Power Syst. Res. 54 (1), 11–17.

Zhou, Z., Hua, C., Chenb, M., Hea, H., Zhang, B., 2010. An improved fuzzy Kalman filter for state estimation of non-linear systems. Int. J. Syst. Sci. 41 (5), 537–546.

Appendix

A Matrix fundamentals

In alphabetical order:

Addition: The addition of two vectors $\mathbf{x}$ and $\mathbf{y}$ with the same dimension n is:

$$\mathbf{x} + \mathbf{y} = \begin{pmatrix} x_1 + y_1 \\ \cdot \\ \cdot \\ \cdot \\ x_n + y_n \end{pmatrix}$$

Vector subtraction is defined in a similar manner. Equivalently, the addition and subtraction of two matrices are defined, when they have identical dimensions.

Column vector: is a vertical array of elements. See further **matrix**.

Condition: The condition of a matrix A denoted by "$\kappa()$" is defined as the product of the norm of a matrix with the norm of the inverse or equivalently as the square root of the ratio of the largest and smallest eigenvalue:

$$\kappa(A) = \|A\| \|A^{-1}\| = \left\{ \lambda_{\max}\left(A^\mathrm{T} A\right) / \lambda_{\min}\left(A^\mathrm{T} A\right) \right\}^{1/2}$$

The condition number is a measure of the numerical inaccuracy encountered in the computation of the inverse of a matrix. As $\kappa(A)$ increases, the matrix is said to be increasingly ill-conditioned.

Derivatives: The matrix-vector differentiation formulas used here are:

$$\frac{d(\mathbf{a}^\mathrm{T}\mathbf{x})}{d\mathbf{x}} = \frac{d(\mathbf{x}^\mathrm{T}\mathbf{a})}{d\mathbf{x}} = \mathbf{a}^\mathrm{T} \quad \frac{d(\mathbf{x}^\mathrm{T} A\mathbf{x})}{d\mathbf{x}} = \mathbf{x}^\mathrm{T}\left(A + A^\mathrm{T}\right) \quad \frac{d^2(\mathbf{x}^\mathrm{T} A\mathbf{x})}{d\mathbf{x}^2} = A + A^\mathrm{T}$$

where $\mathbf{x}$ and $\mathbf{a}$ are n-column vectors and A is a square $n*n$ matrix.

If the dimensions of the matrices X, A, and B match, the partial derivatives for the trace of matrix products resulting in a square matrix are given by:

$$\frac{d[tr\{AXB\}]}{dX} = A^\mathrm{T} B^\mathrm{T} \quad \frac{d[tr\{AX^\mathrm{T} B\}]}{dX} = BA \quad \frac{d[tr\{XAX^\mathrm{T}\}]}{dX} = X\left(A + A^\mathrm{T}\right)$$

Note that the trace derivatives in comparison with the matrix-vector derivatives are defined in the transposed form.

Determinant: To every square $n*n$ matrix A, the determinant representing the volume is a scalar quantity denoted by "$|\ |$". It is defined as the sum of all the products of different signs:

$$|A| = \sum_{i=1}^{n!}(-1)^p A_{1k_1} A_{2k_2} \dots A_{nk_n}$$

where k_1, k_2, ..., k_n run through all possible $n!$ permutations of the numbers $1,2,\dots,n$; p is equal to the number of inversions in each permutation.

If in the matrix A every element of a row (or column) is zero, then the determinant $|A|=0$. Two identical rows or columns also produce a zero determinant. The determinant of the product AB is defined as $|AB|=|A||B|$.

Further be valid the determinant relation $|A^{-1}|=1/|A|$ if $|A| \neq 0$.

Diagonal matrix: A square $n*n$ matrix A is diagonal whenever $A_{ij}=0$ if $i \neq j$ for $i=1,2,\dots,n$ and $j=1,2,\dots,n$.

Eigenvalue: The eigenvalues λ_i, $i=1,2,\dots,n$ of a square $n*n$ matrix A are the roots of the characteristic equation: $|A-\lambda I|=0$. The number of nonzero eigenvalues defines the rank of the matrix A. The determinant $|A|$ equals the product of the eigenvalues, i.e., $|A|=\prod_{i=1}^{n}\lambda_i$ and the trace of A equals $tr\{A\}=\sum_{i=1}^{n}\lambda_i$.

Eigenvector: The n-column eigenvectors $\mathbf{u}_i$, $i=1,2,\dots,n$ are a set of linearly independent or orthonormal vectors associated to the distinct eigenvalues λ_i, which are obtained from the defining relation $A\mathbf{u}_i=\lambda_i \mathbf{u}_i$. Here, A is a square $n*n$ matrix.

Identity matrix: I or I_n is a square $n*n$ matrix with all the diagonal elements unity (one) and all other elements zero.

Inverse matrix: The superscript "-1" denotes the inverse of a matrix. The inverse is only defined for a square $n*n$ matrix A with linear independent rows or columns. The following relation gives the inverse of the matrix A as $AA^{-1}=A^{-1}A=I_n$.

For inversion, the $n*n$ matrix A must have rank n or a nonzero determinant.

Lower triangular matrix: is a square $n*n$ matrix A with $A_{ij}=0$ whenever $j>i$ for $i=1,2,\dots,n$ and $j=1,2,\dots,n$.

Matrix: A matrix is an $m*n$ rectangular array of elements, denoted by a capital letter. The matrix A, consisting of m rows and n columns, is defined as $\{a_{ij}\}$ or $\{A_{ij}\}$, whereas a_{ij} or A_{ij} is the i,jth element for $i=1,2,\dots,m$ and $j=1,2,\dots,n$. An n-column vector is an $n*1$ matrix with n rows and 1 column, an n-row vector an $1*n$ matrix with 1 row and n columns, and a scalar an $1*1$ matrix.

Multiplication: of an $m*n$ matrix A by a scalar b gives the $m*n$ matrix $C=bA=Ab$ composed of elements $C_{ij}=bA_{ij}$ for $i=1,2,\dots,m$ and $j=1,2,\dots,n$.

Nonsingular matrix: A square $n*n$ matrix A is said to be nonsingular if the inverse A^{-1} exists or determinant $|A| \neq 0$ and rank $A=n$. Otherwise A is singular.

Norm: The norm of a vector denoted by "$\| \ \|$" defines a measure of distance or magnitude. Various definitions of the norm are possible. The Euclidian norm is defined as the square root of the sum of squared elements or as the vector product $\|\mathbf{x}\| = (\mathbf{x}^T\mathbf{x})^{1/2}$. The norm of a matrix A used here is $\|A\| = \{\lambda_{max}(A^TA)\}^{1/2}$, where λ_{max} is the largest eigenvalue of the matrix product A^TA.

Null matrix: or zero matrix is an $m*n$ matrix with all the elements zero.

Orthogonal matrix: A square $n*n$ matrix A wherefore its inverse equals the transpose, i.e., $A^{-1} = A^T$. If A is orthogonal, then $A^TA = AA^T = I_n$ and $|A| = \pm 1$.

Orthonormal vectors: A set of vectors $\{\mathbf{x}_1, \mathbf{x}_2, ..., \mathbf{x}_n\}$ wherefore $\mathbf{x}_i^T\mathbf{x}_j = \delta(i,j)$; $\delta(i,j)$ is the Kronecker delta with $\delta(i,j) = 1$ for $i = j$ and $\delta(i,j) = 0$ for $i \neq j$.

Positive definite matrix: If A is a square $n*n$ matrix and $\mathbf{x}$ is an n-column vector, then the matrix is positive definite if the quadratic form $\mathbf{x}^TA\mathbf{x} > 0$ for all $\mathbf{x} \neq 0$.

Product: If $\mathbf{x}$ and $\mathbf{y}$ are n-column vectors with the elements x_i and y_i, respectively, then the vector product $\mathbf{x}^T\mathbf{y} = \sum_{i=1}^{n} x_i y_i$ is a scalar, whereas the product

$$\mathbf{x}\mathbf{y}^T = \begin{pmatrix} x_1y_1 & \cdots & x_1y_n \\ \cdot & & \cdot \\ \cdot & & \cdot \\ \cdot & & \cdot \\ x_ny_1 & \cdots & x_ny_n \end{pmatrix}$$

yields an $n*n$ matrix. The same rules holds for matrix*vector and matrix*matrix products. The $m*l$ matrix product $C = AB$ can be defined only if the number of columns of the $m*n$ matrix A equals the number of rows of the $n*l$ matrix B. Every element of the matrix C is computed as $C_{ij} = \sum_{k=1}^{n} A_{ik}B_{kj}$ for $i = 1, 2, ..., m$ and $j = 1, 2, ..., l$.

Rank: The rank of a matrix is the dimension of the largest square matrix in A formed by deleting rows and columns, which has a nonzero determinant. A square $n*n$ matrix A has rank m with $m \leq n$.

Row vector: is a horizontal array of elements. See further **matrix**.

Scalar: a scalar is defined as a single element with a numerical value and is denoted by a lower-case letter. A scalar is defined as an $1*1$ matrix.

Singular matrix: See **nonsingular matrix**.

Square matrix: a matrix with the same number of rows and columns.

Subtraction: See **addition**.

Symmetric matrix: A square $n*n$ matrix A is symmetric if $A^T = A$.

Trace: The trace of a matrix is designated as "$tr\{\ \}$" and is defined as the sum of the diagonal elements A_{ii} of a square $n*n$ matrix A or: $tr\{A\} = \sum_{i=1}^{n} A_{ii}$.

The following trace identity is valid: $\mathbf{x}^TA\mathbf{x} = tr\{A\mathbf{x}\mathbf{x}^T\} = tr\{\mathbf{x}\mathbf{x}^TA\}$, where $\mathbf{x}$ is an n-column vector.

Transpose: The superscript "T" denotes the transpose of a matrix or a vector. This linear operation can be formulated as interchanging rows and columns.

An n-column vector $\mathbf{x}$ becomes an n-row vector $\mathbf{x}^T$ and vice versa; an $m*n$ matrix A becomes an $n*m$ matrix A^T after transposition.

Unit vector: $\mathbf{e}_i$ is an n-column vector with the ith element unity (one) and all the other elements zero.

Upper triangular matrix: is a square $n*n$ matrix A with $A_{ij}=0$ whenever $i > j$ for $i = 1,2,...,n$ and $j = 1,2,...,n$.

Vector: A vector is an array of elements arranged in a column. It is commonly designated by a bold lower-case letter and referred to as the vector $\mathbf{x}$. The number of elements in the vector is its dimension. x_i denotes the ith element of the vector $\mathbf{x}$.

Zero matrix: See **null matrix**.

B   Statistics fundamentals

B.1   Normal distribution

The continuous distribution having the probability density function:

$$p(x) = \frac{1}{\sigma\sqrt{2\pi}} e^{-\frac{1}{2}\left(\frac{x-\mu}{\sigma}\right)^2}$$

is called the normal distribution or Gaussian distribution. A random variable having this distribution is said to be normal or normally distributed. The probability density function $p(x)$ is a bell-shaped curve and symmetric around the mean μ. The parameter σ is the standard deviation and σ^2 the variance of the distribution.

The cumulative distribution function is given by the integral:

$$P(x) = \frac{1}{\sigma\sqrt{2\pi}} \int_{-\infty}^{x} e^{-\frac{1}{2}\left(\frac{v-\mu}{\sigma}\right)^2} dv$$

This integral cannot be evaluated by elementary methods, but can be expressed by the standardized cumulative distribution function as the integral:

$$\Phi(z) = \frac{1}{\sqrt{2\pi}} \int_{-\infty}^{z} e^{-\frac{1}{2}u^2} du$$

Note that this is the cumulative normal distribution with mean $\mu=0$ and variance $\sigma^2 = 1$. Let us set $u=(v-\mu)/\sigma$. Then $du/dv = 1/\sigma$ and $dv = \sigma du$.

The integration over v from $-\infty$ to x corresponds with the integration over u from $-\infty$ to $(x-\mu)/\sigma$. Thus we obtain:

$$\Phi(z) = \frac{1}{\sigma\sqrt{2\pi}} \int_{-\infty}^{(x-\mu)/\sigma} e^{-\frac{1}{2}u^2} \sigma du$$

σ drops out in this equation. Hence, we have the relationship with $z = (x - \mu)/\sigma$:

$$P(x) = \Phi\left(\frac{x - \mu}{\sigma}\right)$$

The standardized normal distribution is also called the z-distribution and is one of the probability density functions used in statistics for hypothesis testing and determining confidence intervals. The z-test compares two sample means whose variances are known. The critical z-value is employed for determining the confidence interval of the sample mean when the variance is known. It is further used for determining the confidence interval of the elements in the estimated state by recursive least squares, the Kalman filter, or smoothing when the noise variances are known a priori. Critical z-values for a given confidence level α are found in statistical handbooks.

B.2   Chi-square distribution

Let $X_1, X_2, \ldots, X_N$ be N independent normal random variables with mean 0 and variance 1. The sum of their squares is generally denoted by χ^2, that is:

$$\chi^2 = X_1^2 + X_2^2 + \cdots + X_N^2$$

The corresponding probability density function is called the chi-square distribution:

$$p(x) = \frac{1}{2^{n/2}\Gamma(n/2)} x^{(n-2)/2} e^{-x/2}$$

when $x > 0$ and $p(x) = 0$ when $x \le 0$. Here, n is a positive integer and is called the number of degrees of freedom for the distribution. The chi-square distribution has a mean value $\mu = n$ and variance $\sigma^2 = 2n$. For $n = 1$ and $n = 2$, the curves are monotone. For $n > 2$, the curves have a maximum at $x = n - 2$.

The cumulative distribution function is defined by the integral:

$$P(\chi^2) = \frac{1}{2^{n/2}\Gamma(n/2)} \int_0^{\chi^2} u^{(n-2)/2} e^{-u/2} du$$

In both distributions, $\Gamma(\lambda)$ is the so-called gamma function defined by the integral:

$$\Gamma(\lambda) = \int_0^\infty t^{\lambda-1} e^{-t} dt$$

Replacing λ with $\lambda + 1$ and integrating by parts, we have the basic relation $\Gamma(\lambda + 1) = \lambda\Gamma(\lambda)$.

When $\lambda=1$ then $\Gamma(1)=1$, $\Gamma(2)=1\cdot\Gamma(1)=1!$ and $\Gamma(3)=2\cdot\Gamma(2)=2!$.

In general, we have $\Gamma(n+1)=n!$ and when n is even then: $\Gamma(n/2)=(n/2-1)!$.

For the odd case with $n=1$ gives $\Gamma(1/2)=\sqrt{\pi}$, $\Gamma(3/2)=1/2\cdot\Gamma(1/2)=1/2\sqrt{\pi}$ and $\Gamma(5/2)=3/2\cdot\Gamma(3/2)=3/4\sqrt{\pi}$, etc.

The chi-square distribution or χ^2-distribution is a special case of the gamma distribution and is one of the probability density functions used in statistics for hypothesis testing and determining confidence intervals. A chi-square test is used for the goodness of fit of an observed frequency distribution compared with an expected theoretical distribution. Critical χ^2-values are employed for determining the confidence interval of the adapted noise variance in classical least squares or curve fitting. For verification, the standardized innovations or residuals by recursive least squares, the Kalman filter, or smoothing are used in a chi-square test.

Critical table values of chi-square for a given confidence level α and n degrees of freedom can be found in statistical handbooks.

B.3   Fisher distribution

Let χ_1^2 and χ_2^2 be independent random variables both having chi-square distributions with m and n as the numbers of degrees of freedom. Here, m and n are positive integers. Then the random variable: $F=\chi_1^2/m\cdot n/\chi_2^2$ has a probability density function called the Fisher distribution:

$$p(x)=\frac{\Gamma((m+n)/2)}{\Gamma(n/2)\Gamma(m/2)}m^{m/2}n^{n/2}\frac{x^{(m-2)/2}}{(mx+n)^{(m+n)/2}}$$

when $x>0$ and $p(x)=0$ when $x\leq 0$. The beta function $B(m,n)$ is defined by:

$$B(m,n)=\frac{\Gamma(m/2)\Gamma(n/2)}{\Gamma((m+n)/2)}$$

$\Gamma(\lambda)$ is the gamma function as defined in the section of the chi-square distribution.

For $n>2$ the Fisher distribution has a mean value $\mu=n/(n-2)$ and for $n>4$ the variance is:

$$\sigma^2=2\left(\frac{n}{n-2}\right)^2\frac{m+n-2}{m(n-4)}$$

The cumulative distribution function is defined by the integral:

$$P(F)=\frac{\Gamma((m+n)/2)}{\Gamma(n/2)\Gamma(m/2)}m^{m/2}n^{n/2}\int_0^F\frac{u^{(m-2)/2}}{(mu+n)^{(m+n)/2}}du$$

The Fisher distribution is also called the F-distribution. It is a special case of the beta distribution and is one of probability density functions used in statistics for hypothesis testing and determining confidence intervals. It is used for selection by comparing two independent chi-square values. The critical F-value is employed for determining the confidence region of the estimate in classical or recursive least squares, curve fitting, the Kalman filter, or smoothing. An F-test verifies or selects a linear projection in multidimensional space of the estimate and covariance matrix.

If F follows from a F-distribution with (m,n) degrees of freedom and confidence level $1 - \alpha$, then $1/F$ follows from a F-distribution with a confidence level α and (n,m) degrees of freedom, i.e., $F(\alpha,n,m) = 1/F(1 - \alpha,m,n)$.

Critical Fisher F-values for a given confidence level α and (m,n) degrees of freedom can be found in statistical handbooks.

B.4   Student's t distribution

The Student's t distribution is the distribution of the random variable $t = z/\sqrt{\chi^2/n}$, where n is a positive integer that specifies the number of degrees of freedom. z and χ^2 are independent random variables: z has a normal distribution with mean 0 and variance 1, and χ^2 has a chi-square distribution with n degrees of freedom.

The Student's t distribution has a probability density function:

$$p(x) = \frac{\Gamma((n+1)/2)}{\sqrt{n\pi}\Gamma(n/2)} \frac{1}{(1+x^2/n)^{(n+1)/2}}$$

$\Gamma(\lambda)$ is the gamma function as defined in the section of the chi-square distribution.

The probability density function $p(x)$ is a bell-shaped curve and symmetric around zero. For $n = 1$, this is the Cauchy distribution, which has no mean value. For $n > 1$, the distribution function has a mean value of $\mu = 0$. For $n = 1$ and $n = 2$, the Student's t distribution has no variance. For $n > 2$, we obtain the variance $\sigma^2 = n/(n - 2)$.

As n increases, the Student's t distribution approaches the standardized normal distribution with mean $\mu = 0$ and variance $\sigma^2 = 1$.

The cumulative distribution function is given by the integral:

$$P(t) = \frac{\Gamma((n+1)/2)}{\sqrt{n\pi}\Gamma(n/2)} \int\limits_{-\infty}^{t} \frac{1}{(1+u^2/n)^{(n+1)/2}} du$$

The Student's t distribution is also called the t-distribution and is one of the probability density functions used in statistics for hypothesis testing and determining confidence intervals. The t-test compares two sample means whose variances are unknown and adapted. The t-value is employed for determining the confidence interval of the sample mean when the variance is unknown and adapted. It is further used for determining the confidence interval of the elements in the estimate by classical least squares or curve fitting with adaptation of the noise variance.

The Student's t value can be found by the relation $t^2(1 - \alpha/2, n) = F(1 - \alpha, 1, n)$ with a confidence level α and n degrees of freedom.

Critical table values of Student's t for a given confidence level α and n degrees of freedom can be found in statistical handbooks.

C Square root filtering

C.1 Motivation

Although it is theoretically impossible for the covariance matrix to have negative eigenvalues, such a condition can result from numerical computations using finite wordlength in a computer. This may happen when the measurements are very precise or the system becomes nearly unobservable. Such a condition can lead to subsequent divergence or total failure of the Kalman filter. As a solution, square root filters exhibit improved numerical precision and stability, particularly in ill-conditioned systems. Although square root filters are algebraically equivalent to the Kalman filter, the same precision can be achieved with approximately half the wordlength in a computer; moreover, these methods are completely successful in maintaining the positive definiteness of the covariance matrix. These good numerical characteristics, combined with modest additional computation time and memory storage requirements, make the square root approach a valuable alternative in many applications.

C.2 Cholesky decomposition

Let A be an $n*n$ symmetric matrix. Then there exists at least one $n*n$ square root matrix $\sqrt{A}$ such that $\sqrt{A}\sqrt{A}^T = A$. In fact, there are in general many matrices, which satisfies the earlier equation. The essential idea is to replace the covariance matrix P for its square root matrix $\sqrt{P}$ and to compute the estimated state using an optimal gain calculated in terms of $\sqrt{P}$ instead of P itself.

The basic idea is that before and after the measurement update at index k $S(k/k-1)$ and $S(k/k)$ can be defined as:

$$S(k/k-1)S(k/k-1)^T = P(k/k-1) \quad \text{and} \quad S(k/k)S(k/k)^T = P(k/k)$$

The square root covariance matrices are not uniquely defined and square root filters can be expressed in terms of general square root matrices. Any symmetric positive definite matrix can be factored by the Cholesky decomposition into the product of a lower triangular matrix and its transpose. Although $\sqrt{A}$ is not uniquely defined, a unique lower triangular Cholesky square root matrix can be $\sqrt[c]{A}$ such that $\sqrt[c]{A}\sqrt[c]{A}^{\mathrm{T}} = A$.

The elements of the Cholesky square root matrix $\sqrt[c]{A}$ can be generated sequentially row by row from the left to the right within each row for $i = 1,2,\ldots,n$:

$$
\sqrt[c]{A_{ij}} = \begin{cases}
1/\sqrt[c]{A_{jj}}\left[A_{ij} - \displaystyle\sum_{k=1}^{j-1}\sqrt[c]{A_{ik}}\sqrt[c]{A_{jk}}\right] & j = 1,2,\ldots,i-1 \\[2mm]
\left(A_{ii} - \displaystyle\sum_{k=1}^{i-1}\sqrt[c]{A_{ik}^2}\right)^{1/2} & j = i \\[2mm]
0 & j > i
\end{cases}
$$

In the Carlson filter described later, we use an upper triangular Cholesky square root matrix $\sqrt[cT]{A}$, where the computation runs backward for $j = n$, $n-1,\ldots,1$:

$$
\sqrt[cT]{A_{ij}} = \begin{cases}
0 & i > j \\[2mm]
\left(A_{jj} - \displaystyle\sum_{k=j+1}^{n}\sqrt[cT]{A_{jk}^2}\right)^{1/2} & i = j \\[2mm]
1/\sqrt[cT]{A_{jj}}\left[A_{ij} - \displaystyle\sum_{k=j+1}^{n}\sqrt[cT]{A_{ik}}\sqrt[cT]{A_{jk}}\right] & i = j-1, j-2,\ldots,1
\end{cases}
$$

Thus, the elements of the Cholesky square root matrix $\sqrt[cT]{A}$ are generated sequentially column by column from the bottom to the top within each column.

C.3   Potter filter

Potter developed a square root algorithm in which there was no system noise in the state space model, i.e., $Q(k-1) = 0$. In this case, the prediction in the Kalman filter is:

$$\hat{\mathbf{x}}(k/k-1) = F(k,k-1)\hat{\mathbf{x}}(k-1/k-1)$$

$$P(k/k-1) = F(k,k-1)P(k-1/k-1)F^{\mathrm{T}}(k,k-1)$$

By letting : $P(k-1/k-1) = S(k-1/k-1)S^{\mathrm{T}}(k-1/k-1)$

and : $P(k/k-1) = S(k/k-1)S^{\mathrm{T}}(k/k-1)$

this equation can be rewritten as:

$$S(k/k-1)S^{\mathrm{T}}(k/k-1) = F(k,k-1)S(k-1/k-1)S^{\mathrm{T}}(k-1/k-1)F^{\mathrm{T}}(k,k-1)$$

From this, the index propagation by prediction for the square root filter becomes:

$$\hat{x}(k/k-1) = F(k, k-1)\hat{x}(k-1/k-1)$$

$$S(k/k-1) = F(k, k-1)S(k-1/k-1)$$

The state estimate and the measurement update by filtering is given by:

$$\mathbf{a}(k) = S^{\mathrm{T}}(k/k-1)\mathbf{h}(k)$$

$$b(k) = 1/\left[\mathbf{a}^{\mathrm{T}}(k)\mathbf{a}(k) + R(k)\right]$$

$$\gamma(k) = 1/\left[1 + \{b(k)R(k)\}^{1/2}\right]$$

$$\mathbf{k}(k) = b(k)S(k/k-1)\mathbf{a}(k)$$

$$\hat{x}(k/k) = \hat{x}(k/k-1) + \mathbf{k}(k)\{z(k) - \mathbf{h}^{\mathrm{T}}(k)\hat{x}(k/k-1)\}$$

$$S(k/k) = S(k/k-1) - \gamma(k)\mathbf{k}(k)\mathbf{a}^{\mathrm{T}}(k)$$

Note that even $S(k/k-1)$ is a lower triangular matrix $S(k/k)$ will generally not be when this updated form is used. A method that preserves the triangular nature of the square root covariance matrix will be discussed later.

C.4 Carlson filter

If driving system noise enters the state space model, i.e., $Q(k-1) \neq 0$, the Kalman filter propagates from the index $k-1$ to k by prediction:

$$P(k/k-1) = F(k, k-1)P(k-1/k-1)F^{\mathrm{T}}(k, k-1) + Q(k-1)$$

Now we wish to develop an analogous recursion to yield $S(k/k-1)$ in terms of $S(k-1/k-1)$. One means of achieving the desired result is denoted as the matrix RSS (root-sum-square) method:

$$X(k/k-1) = F(k, k-1)S(k-1/k-1)$$

$$P(k/k-1) = X(k/k-1)X^{\mathrm{T}}(k/k-1) + Q(k-1)$$

$$S(k/k-1) = \sqrt[c]{P(k/k-1)}$$

This method actually computes $P(k/k-1)$ and then generates $S(k/k-1)$ as its lower triangular Cholesky square root matrix. Nevertheless, it is almost always the measurement update by filtering and not the index propagation by prediction in the Kalman filter that causes numerical problems in the computations and the matrix RSS method may well be acceptable in many applications.

Two other methods for prediction known as triangular algorithms are the Gram-Schmidt orthogonalization and the Householder technique. These methods are not discussed here because they are rather complex and complicated.

One significant drawback of the square root covariance filter by Potter is that the triangularity of the square root matrix is destroyed during the measurement update.

In a more recent algorithm, the Carlson filter keeps the square root covariance matrix always as an upper triangular matrix.

Setting the scalar d_0 and the n-column vectors $\mathbf{e}_0$ and $\mathbf{a}$ initializes the Carlson filter as $d_0 = R(k)$, $\mathbf{e}_0 = 0$, and $\mathbf{a} = S^T(k/k-1)\mathbf{h}(k)$ iterating for $i = 1,2,\ldots,n$:

$$d_i = d_{i-1} + a_i^2$$

$$b_i = (d_{i-1}/d_i)^{1/2}$$

$$c_i = a_i/(d_{i-1}d_i)^{1/2}$$

$$\mathbf{e}_i = \mathbf{e}_{i-1} + S_i^- a_i$$

$$S_i^+ = S_i^- b_i - \mathbf{e}_{i-1} c_i$$

In the recursion, S_i^- denotes the ith column of $S(k/k-1)$ and both it and $\mathbf{e}_i$ consists of zeros below the ith element. After n iterations, $S(k/k)$ is produced as: $S(k/k) = [S_1^+ \; S_2^+ \; \ldots \; S_n^+]$, which is an upper triangular matrix.

The estimated state update with the gain vector $\mathbf{k}(k) = \mathbf{e}_n/d_n$ is then given by:

$$\hat{\mathbf{x}}(k/k) = \hat{\mathbf{x}}(k/k-1) + \mathbf{k}(k)\{z(k) - \mathbf{h}^T(k)\hat{\mathbf{x}}(k/k-1)\}$$

For index propagation by prediction, Carlson suggested the matrix RSS method, but now with an upper triangular Cholesky square root matrix as generated before in the Potter filter replacing the lower triangular matrix.

C.5   *U-D* filter

Another approach of the optimal filter algorithm is known as *U-D* factorization developed by Bierman and Thornton. Instead of decomposing the covariance matrix into a square root matrix as described before, this method expresses the covariance matrices before and after measurement incorporation by:

$$P(k/k-1) = U(k/k-1)D(k/k-1)U^T(k/k-1)$$

$$P(k/k) = U(k/k)D(k/k)U^T(k/k),$$

where the U-matrices are upper triangular and unitary (with ones along the diagonal) and the D-matrices are diagonal. Although square root covariance matrices are never explicitly evaluated by this method, $UD^{1/2}$ corresponds directly to the matrix S in the Carlson filter. The matrices U and D are not unique and they can be generated through an algorithm closely related to the upper triangular Cholesky decomposition. Let the $n*n$ covariance matrix P be symmetric and positive definite.

For the columns running backward, from the nth column to the first and from the bottom to the top, compute for $j = n,\ n-1,\ldots,1$:

$$D_{jj} = P_{jj} - \sum_{k=j+1}^{n} D_{kk} U_{jk}^2$$

$$U_{ij} = \begin{cases} 0 & i > j \\ 1 & i = j \\ \left[P_{ij} - \displaystyle\sum_{k=j+1}^{n} D_{kk} U_{ik} U_{jk} \right] / D_{jj} & i = j-1, j-2, \ldots, 1 \end{cases}$$

The measurement update by filtering for the U-D factorization can be specified. $U(k/k-1)$ and $D(k/k-1)$ are available from a previous index propagation by prediction. One computes:

$$\mathbf{f} = U^{\mathrm{T}}(k/k-1)\mathbf{h}(k)$$

$$v_j = D_{jj}(k/k-1)f_j \quad j = 1, 2, \ldots, n$$

$$a_0 = R(k)$$

Then for $i = 1, 2, \ldots, n$ calculate:

$$a_i = a_{i-1} + f_i v_i$$

$$D_{ii}(k/k) = D_{ii}(k/k-1)a_{i-1}/a_i$$

$$b_i \leftarrow v_i$$

$$p_i = -f_i/a_{i-1}$$

$$\left. \begin{array}{l} U_{ji}(k/k) = U_{ji}(k/k-1) + b_j p_i \\ b_j \leftarrow b_j + U_{ji}(k/k-1)v_i \end{array} \right\} \quad j = 1, 2, \ldots, i-1$$

$\leftarrow$ denotes replacement exploiting the technique of writing over old variables for efficiency. The gain vector $\mathbf{k}(k)$ can be calculated in terms of the n-column vector $\mathbf{b}$ made up of the components $b_1, b_2, \ldots, b_n$ computed in last iteration of the equation above. Then the estimated state update with the gain vector $\mathbf{k}(k) = \mathbf{b}/a_n$ becomes:

$$\hat{\mathbf{x}}(k/k) = \hat{\mathbf{x}}(k/k-1) + \mathbf{k}(k)\{z(k) - \mathbf{h}^{\mathrm{T}}(k)\hat{\mathbf{x}}(k/k-1)\}$$

For the index propagation by prediction of the U-D factors, the following matrix RSS method can be employed:

$$P(k-1/k-1) = U(k-1/k-1)D(k-1/k-1)U^{\mathrm{T}}(k-1/k-1)$$

$$P(k/k-1) = F(k, k-1)P(k-1/k-1)F^{\mathrm{T}}(k, k-1) + Q(k-1)$$

$$U(k/k-1)D(k/k-1)U^{\mathrm{T}}(k/k-1) = P(k/k-1)$$

For prediction, the U-D matrices can also be processed by a Gram-Schmidt orthogonalization, whose algorithm is not discussed here.

D   Chemical reaction networks

The kinetic equations that describe chemical reactions are modeled on the basis of mass-action kinetics. The formulation may be illustrated using an example of a reversible reaction:

$$K_2SO_4 + CaCl_2 \underset{k2}{\overset{k1}{\rightleftarrows}} CaSO_4 + 2KCl,$$

where k_1 and k_2 are the reaction rates for the forward and backward reactions, respectively. Let X_1, X_2, X_3, X_4, denote the species K_2SO_4, $CaCl_2$, $CaSO_4$, KCl, respectively, so the equation may be rewritten as:

$$X_1 + X_2 \overset{k_1}{\rightarrow} X_3 + 2X_4$$

$$X_3 + 2X_4 \overset{k_2}{\rightarrow} X_1 + X_2$$

In these expressions there are four species ($s=4$), so that a vector of species may be constructed: $X = [X_1 \ X_2 \ X_3 \ X_4]^T$. Similarly a vector of rates may also be constructed, where $k = [k_1 \ k_2]^T$. On the basis of stoichiometric coefficients, one can write

$$A = \begin{bmatrix} 1 & 1 & 0 & 0 \\ 0 & 0 & 1 & 2 \end{bmatrix}, \quad B = \begin{bmatrix} 0 & 0 & 1 & 2 \\ 1 & 1 & 0 & 0 \end{bmatrix}$$

and in the more general case with s species $X_1, \ldots, X_s$, and r reactions comprising a reaction network,

$$\underbrace{\sum_{j=1}^{s} A_{ij}X_j}_{\text{reactants}} \overset{k_i}{\rightarrow} \underbrace{\sum_{j=1}^{s} B_{ij}X_j}_{\text{products}}, \quad i=1,\ldots,r$$

In the given description, reversible reactions are modeled as separate irreversible reactions and stoichiometric coefficients A_{ij} and B_{ij} are assumed to be non-negative integers. Generally, $X = [X_1 \ X_2 \ \ldots \ X_s]^T$, $k = [k_1 \ \ldots \ k_r]^T \in [0, \infty)^r$ and A and B denote the $r \times s$ non-negative matrices $A = [A_{ij}]$ and $B = [B_{ij}]$. Each species appears in the reaction network with at least one nonzero coefficient so that none of the columns of $[A \ B]^T$ is zero. The kinetic equations of the reaction are as follows:

$$\begin{aligned}
\dot{x}_1(t) &= -k_1 x_1(t)x_2(t) + k_2 x_3(t)x_4^2(t) & x_1(0) &= x_{10} \quad t \geq 0 \\
\dot{x}_2(t) &= -k_1 x_1(t)x_2(t) + k_2 x_3(t)x_4^2(t) & x_2(0) &= x_{20} \\
\dot{x}_3(t) &= k_1 x_1(t)x_2(t) - k_2 x_3(t)x_4^2(t) & x_3(0) &= x_{30} \\
\dot{x}_4(t) &= 2k_1 x_1(t)x_2(t) - 2k_2 x_3(t)x_4^2(t) & x_4(0) &= x_{40}
\end{aligned}$$

In the more general case, one can write:

$$\dot{x}(t) = (B - A)^T \left[k \circ x^A(t) \right] \quad x(0) = x_0 \quad t \geq 0,$$

where the notation $k \circ x^A$ denotes vector matrix exponentiation (component by component and entry-by-entry multiplication). A comprehensive discussion on the realizability and reducibility of the kinetic equations as well as on the stability analysis of equilibria can be found in Chellaboina et al. (2009). That work also considers the realizability problem, which is concerned with the inverse problem of constructing a reaction network having specified essentially non-negative dynamics. A practical application of realizability theory is in the identification of possible reaction networks in biochemical studies.

Reference

Chellaboina, V., Bhat, S.P., Haddad, W.M., Bernstein, D.S., 2009. IEEE Control Systems Magazine 60–78.

Bibliography

Adams, M.J., 2004. Chemometrics in Analytical Spectroscopy. RSC Press.

Anderson, B.D.O., More, J.B., 1979. Optimal Filtering. Prentice Hall.

Beebe, K.R., Pell, R.J., Seasholtz, M.B., 1998. Chemometrics: A Practical Guide. Wiley.

Bozic, S.M., 1979. Digital and Kalman Filtering. Edward Arnold Publishers.

Brammer, K., Siffling, G., 1975. Kalman-Bucy-Filter. R. Oldenburg Verlag.

Brereton, R.G., 1990. Chemometrics: Applications of Mathematics and Statistics to Laboratory Systems. Ellis Horwood.

Brereton, R.G., 2003. Chemometrics: Data Analysis for the Laboratory and Chemical Plant. Wiley.

Brereton, R.G., 2007. Applied Chemometrics for Scientists. Wiley.

Brown, R.G., 1983. Introduction to Random Signal Analysis and Kalman Filtering. Wiley.

Brown, S.D., Tauler, R., Walczak, B. (Eds.), 2009. Comprehensive Chemometrics. In: vols. 1–4. Elsevier.

Bruns, R.E., Scarminio, I.S., de Barros Neto, B., 2006. Statistical Design Chemometrics. Elsevier.

Candy, J.V., 1986. Signal Processing: The Model-Based Approach. McGraw-Hill.

Chau, F., Liang, Y., Gao, J., Shao, X., 2004. Chemometrics: From Basics to Wavelet Transform. Wiley.

Chui, C.K., Chen, G., 1987. Kalman Filtering. Springer-Verlag.

Eubank, R.L., 2005. A Kalman Filter Primer. Chapman & Hall/CRC.

Federov, V.V., 1972. Theory of Optimal Experiments. Academic Press.

Gelb, A. (Ed.), 1974. Applied Optimal Estimation. MIT Press.

Gemperline, P. (Ed.), 2006. Practical Guide to Chemometrics, second ed. CRC Press.

Golub, G.H., Van Loan, C.F., 1983. Matrix Computations. North Oxford Academic.

Graupe, D., 1972. Identification of Systems. Robert E Krieger Publishing.

Grewal, M.S., Andrews, A.P., 2001. Kalman Filtering: Theory and Practice. Prentice Hall.

Hecht, H.G., 1993. Mathematics in Chemistry. Prentice Hall.

Kateman, G., Buydens, L.M.C., 1993. Quality Control in Analytical Chemistry, second ed. Wiley.

Kowalski, B.R. (Ed.), 1977. Chemometrics: Theory and Application. American Chemical Society.

Kramer, R., 1998. Chemometric Techniques for Quantitative Analysis. Marcel Dekker.

Krebs, V., 1980. Nichtlineare Filterung. R. Oldenburg Verlag.

Kreyszig, E., 1970. Introductionary Mathematical Statistics. Wiley.

Liteanu, C., Rica, I., 1980. Statistical Theory and Methodology of Trace Analysis. Ellis Horwood.

Malinowski, E.R., 2002. Factor Analysis in Chemistry. Wiley.

Mark, H., Workman, J., 2003. Statistics in Spectroscopy. Elsevier.

Mark, H., Workman, J., 2007. Chemometrics in Spectroscopy. Academic Press.

Martens, H., Naes, T., 1989. Multivariate Calibration. Wiley.

Massart, D.L., Dijkstra, A., Kaufmann, L., 1978. Evaluation and Optimization of Laboratory Methods and Analytical Procedures. Elsevier.

Massart, D.L., Vandeginste, B.G.M., Deming, S.N., Michotte, Y., Kaufmann, L., 1988. Chemometrics: A Textbook. Elsevier.

Massart, D.L., Vandeginste, B.G.M., Buydens, L.M.C., de Jong, S., Lewi, P.J., Smeyers-Verbeke, J., 1998. Handbook of Chemometrics and Qualimetrics, Part A and B. Elsevier.

Maybeck, P.S., 1979. Stochastic Models, Estimation and Control. vols. 1–3 Academic Press.

Mendel, J.M., 1973. Discrete Techniques of Parameter Estimation. Marcel Dekker.

Miller, J.C., Miller, J.N., 1988. Statistics for Analytical Chemistry, second ed. Ellis Horwood.

Miller, J.N., Miller, J.C., 2005. Statistics and Chemometrics for Analytical Chemistry, fifth ed. Pearson Education.

Morgan, E., 1991. Chemometrics: Experimental Design. Wiley.

Norton, J.P., 1986. An Introduction to Identification. Academic Press.

Oja, E., 1983. Subspace Methods of Pattern Recognition. Wiley.

Otto, M., 1999. Chemometrics: Statistics and Computer Application in Analytical Chemistry. Wiley.

Poulisse, H.N.J., 1983. State- and Parameter Estimation as Chemometrical Concepts. Ph. D. Thesis, Radboud University Nijmegen, The Netherlands.

Schaum Outline Series, 1974. Matrices. McGraw-Hill.

Scheeren, P.J.H., 1988. Modeling, Dataprocessing and Signal Estimation in Analytical Chemistry. Ph. D. Thesis, University of Amsterdam, The Netherlands.

Sharaf, M.A., Illman, D.L., Kowalski, B.R., 1986. Chemometrics. Wiley.

Simon, D., 2006. Optimal State Estimation. Wiley.

Sorenson, H.W., 1980. Parameter Estimation. Marcel Dekker.

Thijssen, P.C., 1986. State Estimation in Chemometrics. Ph. D. Thesis, Radboud University Nijmegen/University of Amsterdam, The Netherlands.

Varmuza, K., Filzmoser, P., 2009. Multivariate Statistical Analysis in Chemometrics. CRC Press.

Young, P., 1984. Recursive Estimation and Time-Series Analysis. Springer-Verlag.

Zijp, W.L., 1974. Handleiding voor Statistische Toetsen. Tjeenk Willink.

Index

Note: Page numbers followed by *f* indicate figures and *t* indicate tables.

Printed in the United States
By Bookmasters